T0321647

INTERNATIONAL YOUNG PHYSICISTS' TOURNAMENT

Problems & Solutions
2015

INTERNATIONAL YOUNG PHYSICISTS' TOURNAMENT

Problems & Solutions
2015

Editors

Sihui Wang
Wenli Gao

Nanjing University, China

World Scientific

EW JERSEY · LONDON · SINGAPORE · BEIJING · SHANGHAI · HONG KONG · TAIPEI · CHENNAI · TOKYO

Published by

World Scientific Publishing Co. Pte. Ltd.

5 Toh Tuck Link, Singapore 596224

USA office: 27 Warren Street, Suite 401-402, Hackensack, NJ 07601

UK office: 57 Shelton Street, Covent Garden, London WC2H 9HE

British Library Cataloguing-in-Publication Data
A catalogue record for this book is available from the British Library.

INTERNATIONAL YOUNG PHYSICISTS' TOURNAMENT
Problems and Solutions 2015

ISBN 978-981-3225-91-6 (pbk)

For any available supplementary material, please visit
http://www.worldscientific.com/worldscibooks/10.1142/10596#t=suppl

Desk Editor: Christopher Teo

Typeset by Stallion Press
Email: enquiries@stallionpress.com

Printed in Singapore

Dedication and Acknowledgement

This book is dedicated to our student laboratory 326, a place where we shed tears and sweat, where we shared sleepless nights and moments of eureka!

The authors are also grateful to the support from School of Physics, Nanjing University and Jiangsu Physical Society.

Contents

Chapter 1

2015 Problem 3: Artificial Muscles

Xin Yuan[1]*, Ruyang Sun[2]†, Wenli Gao[3], Huijun Zhou[3]

[1] *School of Chemistry and Chemical Engineering, Nanjing University*
[2] *Kuang Yaming Honors School, Nanjing University*
[3] *School of Physics, Nanjing University*

In this solution, we investigate the thermally-actuated artificial muscles made from fishing lines by twist insertion. We give an explanation about the fabrication of the muscle, and illuminate the influences of several parameters during the fabrication process. Based on the properties of polymer, a model is put forward to describe the observed torsional actuation of twisted fiber. A relation between fiber's torsional actuation and muscle's tensile actuation is deduced, which proves that the spring-like structure of the muscle plays an important role in the thermal contraction.

1. Introduction

Attach a polymer fishing line to an electric drill and apply tension to the line. As it twists, the fibre will form tight coils in a spring-like arrangement. Apply heat to the coils to permanently fix that spring-like shape. When you apply heat again, the coil will contract. Investigate this artificial muscle.

The term "artificial muscle" is generally used to describe materials or devices that can reversibly contract, expand, or rotate within one component due to an external stimulus (such as voltage, current, pressure or temperature). The high flexibility, versatility and power-to-weight ratio of this kind of materials, compared with traditional rigid actuators, hint that they have the potential to be a highly disruptive emerging technology.[1]

Haines *et al.*, first reported this kind of materials in 2014.[2] They fabricated artificial muscles from commercial fishing lines, using a common

*E-mail: 141130130@smail.nju.edu.cn
†E-mail: 141242053@smail.nju.edu.cn

1

electrical drill. Some investigations of relevant properties and an explanation of the actuation were also given.

In this solution, we mainly investigate the fabrication and actuation of the artificial muscles based on their work. In the fabrication part, we investigate under what condition the artificial muscle will form, and how the relevant parameters influence the structure of muscle. In the actuation part, we illuminate how the muscle contracts if heated, based on the properties of polymer and observation.

2. Fabrication of Muscles

2.1. *Preliminary Experiment*

In preliminary experiment, we fix one end of the fishing line, and hold the other end in hand to control the tension. We find that if we keep too large tension, the fiber will break; if we keep too little tension, it will snarl; only a suitable tension will lead to spring-like muscle. And as we change the tension while we fabricate the muscle, the muscle will have different thickness. Also, if we change the spin rate of the drill like on-off-on, we cannot obtain a homogeneous muscle. All these phenomena indicate that the applied tension and the spin rate should be kept constant during the fabrication.

2.2. *Manufacture*

After several preliminary experiments, it can be found that constant twist insertion rate and tension inside the fiber could ensure the muscles to be uniform and homogeneous. Based on these, a feasible apparatus is designed as shown in Fig. 1: One end of the fishing line is clumped into an electric drill chunk, while the other end is tied to a suitable load. With the help of a vertical pole and a horizontal rod, the lower end of the line is constrained from rotating.

During the fabrication, as the friction between the rod and the pole could be neglected compared to the unchanged weight of the rod and load, the tension inside the fishing line could be assumed to be constant. The common electric drill provides a constant twist insertion rate (about 8000 rpm).

When the drill begins to work[‡], it can be observed that the load is first

---

[‡]See supplementary materials, Fabrication Process, video 1.

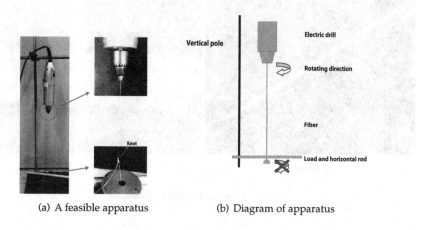

(a) A feasible apparatus   (b) Diagram of apparatus

Fig. 1.   Apparatus.

lifted up and the line begins to coil up from the knot in the lower end. The coil grows all through the whole line quickly, leading to a notable decrease of the distance between the two ends of the line. Then we take the coiled line out, anneal the line at about 50 °C. During annealing, the coiled line will shrink with an increase of the diameter. About one minute later, the spring-like shape will be fixed permanently and an artificial muscles is obtained (Fig. 9). Actually, if we put the newly prepared coiled line aside for about several hours, it will also undergo the same procedure of line shrink and diameter increase, and finally stabilize to almost the same size and shape as those of the annealed one. This comparison indicates that annealing process expedites the relief of residual stress, which will spontaneously occur over a long time at room temperature.

In order to find out the suitable range of load weight for our fishing line, different loads are applied to manufacture the artificial muscles. As to our sample, a nylon fiber with a diameter of 0.46 millimeters, the suitable load range is from about 160 grams to 310 grams.

Within the range, different weights result in muscles of different diameters. The less load results in larger diameter, which means a higher spring index (the ratio of muscle diameter to the fiber diameter), and larger load results in smaller diameter. The data is shown in Fig. 3.

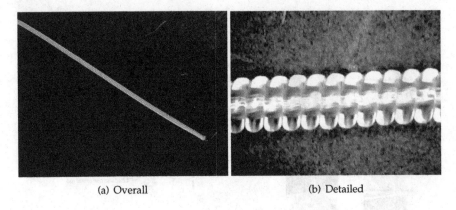

(a) Overall   (b) Detailed

Fig. 2.   Fabricated artificial muscle.

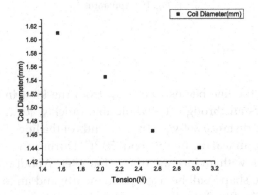

Fig. 3.   Relation between the applied load and corresponding coil diameter.

2.3. *Analysis*

Twist insertion directly leads to torsion inside the fishing line, which helps the muscle maintain its helical structure under tension. As shown in Fig. 4(a), we take a cross section A to divide the fiber into two parts and take a look at the upper one. From mechanics of materials, we know there is an internal force F_a applied on the cross section, equal to the tension F applied to the upper part but directing to the contrary. Then the two forces generate an internal torque T_a acting on cross section A to counteract it. We decompose the torque T_a into two parts, T_{a1} vertical to the section and T_{a2} parallel to the section. T_{a1} is due to torsion of the fiber, and T_{a2} is the bending moment. Defining α_c as the bias angle of fiber relative to the transverse section of the muscle, we will get $T_{a1} = T_a \cos \alpha_c$. By taking a cross section B, analysing the lower part of it, F_b, T_b, T_b1 and T_b2 can be obtained similarly. Due to the symmetry of the helix, the torque on A should have the same value on B, with a different direction. So we use the torque value on section A for convenience.

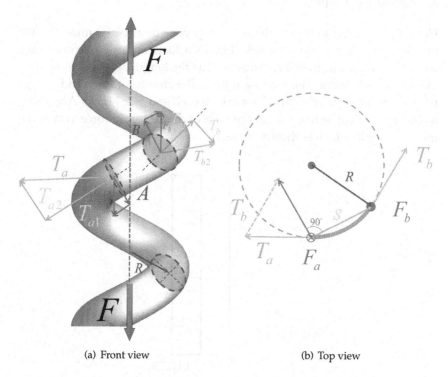

(a) Front view   (b) Top view

Fig. 4. Diagram of helical structure of artificial muscle.

Then we analyse the fiber between section A and B (see the top view by Fig. 4(b)). Assuming the distance between the two sections as s, the torque due to internal forces F_a and F_b is $T = Fs$. Since torques T_a and T_b are horizontal, the resultant torque is just equal to $\frac{s}{R} T_a$, where R is the mean radius of muscle. The taken out part is in balance, so $Fs = \frac{s}{R} T_a$, and we get the relation between the torque of torsion T_{a1} and tension F:

$$T_{a1} = FR \cos \alpha_c \qquad (1)$$

As twist is inserted to the fishing line, the torque inside the fiber increases due to increasing torsion. Because the tension F is kept constant, according to Eq. (1), increasing torque leads to decreasing bias angle α_c. Since the fabricated muscle already has a minimum bias angle (One round contacts the other tightly), the upper end of the muscle coils up to decrease the α_c there and we can see the muscle "growing".

3. Mechanical Property

Here, we investigate the mechanical properties of artificial muscles. We use an artificial muscle to slowly lift up a load on an electronic scale, recording the reading on display as well as the length of whole muscle by video. In each frame, the reading reflects the change of force and we get the length of the muscle with the mark on each end (See Fig. 5). According to our analysis of video, we get the data of the restoring force versus the muscle length, which is shown in Fig. 6.

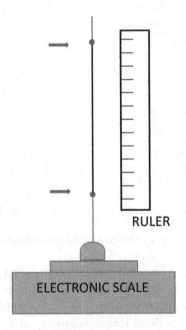

Fig. 5.   Diagram of apparatus in investigating spring constant.

Fig. 6 demonstrates that the spring constant could be divided into two parts. In the first part, the elongation is small and the spring constant is large. As elongation increases to the second part, the spring constant

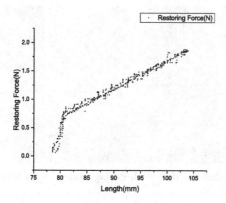

Fig. 6.   Data from a same muscle stretching and contracting several times.

reduces. This piecewise spring constant occurs as the elongation changes, when elongation is small, the coils contact with each other and press adjacent coils, causing the whole muscle performs rigid; when elongation is large enough to make the coils separate, such press disappears, causing the muscle to perform like a common spring.

Comparing different muscles, we find that the muscles fabricated under the same conditions (diameter, load) have a similar spring constant for the same muscle length. If the applied load during fabrication becomes larger while other parameters remain the same, the spring constant per length of the muscle will get correspondingly larger.

4. Thermal Actuation of Muscles

It is obvious that the loaded muscle could be stretched out like a spring. When it is heated under load, a reversible contraction by nearly 10% is observed from 20°C to 80°C. As a vivid example, the movement of the lower end (red mark) of the muscle during the thermal actuation is shown in Fig. 7.[§] With upper end fixed, when the muscle is heated, it shrinks with the lower end moving upward. After it cools down, the lower end goes back to its original place. Here we give a mechanism to explain why this thermal actuated contraction can occur, based on the inherent properties of twisted nylon fishing lines and muscle's helical structure.

[§]See supplementary materials, Thermal Contraction, video 2.

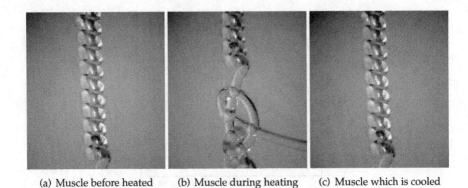

(a) Muscle before heated   (b) Muscle during heating   (c) Muscle which is cooled

Fig. 7.   Lower end of the muscle during the thermal actuation.

4.1. *Rubber Elasticity*

Nylon, as well as polyethylene shown in Haines's work,[2] differs from other materials because it is linear polymer, which means it consists of long linear molecule chains. These chains enable nylon to perform rubber elasticity[3] (entropic elasticity), which depends on the temperature. Here we use ideal chain (freely-jointed chain) model[4] to deduce the relation between rubber elasticity and temperature. Although it is simple, its generality gives us some insights about the physics of polymers.

Ideal chain model assumes a free-jointed polymer molecule chain whose chemical bonds are rigid rods and completely randomly oriented, independent of the orientations and positions of neighbouring bonds and monomers, like a random walk track. The polymer without branch can fit the model better, like nylon and linear polyethylene.

Assuming that the distance b between two monomers is a step of the random walk (Fig. 8), which is the bond length between neighbouring monomers, the total step number N is the number of bonds. Then we can find that if we have a definite fixed total end to end vector $\mathbf{R}$ of an ideal chain, the micro-states number Ω of the chain equals the number of ways from one end to another by random walk, associated to the definite $\mathbf{R}$ and N.

The entropy of the chain is thus equal to

$$S(N, \mathbf{R}) = k \ln \Omega(N, \mathbf{R}) \tag{2}$$

According to the random walk theory, in three dimensions, the $\mathbf{R}$

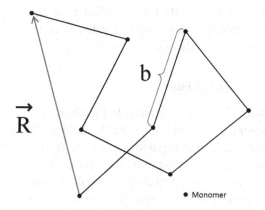

Fig. 8.   Ideal chain and random walk track.

distributes based on the following probability density function:

$$P_{3d}(N, \mathbf{R}) = \left(\frac{3}{2\pi Nb^2} \right)^{3/2} \exp \left(-\frac{3\mathbf{R}^2}{2Nb^2} \right) \tag{3}$$

We know that probability density P_{3d} is defined as

$$P_{3d}(N, \mathbf{R}) = \frac{\Omega(N, \mathbf{R})}{\int \Omega(N, \mathbf{R})d\mathbf{R}} \tag{4}$$

Combined with Eq. (3) and Eq. (4), Eq. (2) could be expressed as,

$$S(N, \mathbf{R}) = -\frac{3}{2}k\frac{\mathbf{R}^2}{Nb^2} + \frac{3}{2}k\ln \left(\frac{3}{2\pi Nb^2} \right) + k\ln \left[\int \Omega(N, \mathbf{R})d\mathbf{R} \right] \tag{5}$$

For a given chain with constant N, the last two terms of Eq. (5) are invariants. The first term tells that when the distance between two ends ($\mathbf{R}$) reduces, the entropy of the chain increases.

By Helmholtz free energy $F = U - TS$, since we neglect any kind of interactions among monomers and chains, the restoring force is given by

$$\mathbf{f} = \frac{\partial F}{\partial \mathbf{R}} = -T\frac{\partial S(N, \mathbf{R})}{\partial \mathbf{R}} = \frac{3kT}{Nb^2}\mathbf{R} \tag{6}$$

Note that the coefficient $\frac{3kT}{Nb^2}$ is temperature dependent, positive correlated to temperature. When temperature increases, this coefficient of $\frac{3kT}{Nb^2}$ will rise. If the restoring force $\mathbf{f}$ is kept constant, the length of the chain will decrease. Under a constant restoring force, an increase temperature will cause the muscle to shrink.

Although this is a formula only applied to a single ideal chain in the microscopic domain, the similar tendency can also apply to the macroscopic in terms of statistics.

4.2. *Actuation of Twisted Fiber*

In our experiment, if we insert twists into the fiber and stop before it coils, we will get a twisted fiber under load. By the mark on the fiber surface, we observe that when such fiber is partly heated, the heated part will untwist, and the unheated part will twist more[¶]. It shows that a twisted fiber will show a tendency to untwist if heated. This phenomenon is discussed in this subsection, and how this property is related to the actuation of the muscle will be discussed in the next one.

The high strength of fishing lines usually benefits from its high degree of polymer alignment along the fiber axis. Based on this and the rubber elasticity of polymer, a shell-like model[2] is given to describe the actuation of twisted fiber qualitatively.

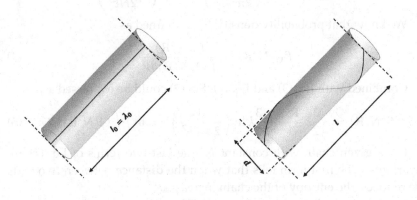

(a) Untwisted fiber and straight chain (subscript "0" identifying the initial state)

(b) Twisted fiber and helical chain

Fig. 9. A molecule chain on a cylindrical shell in shell-like model.

In this model, all the molecule chains are assumed to align along the

---

[¶]See supplementary materials, Thermal actuation of twisted fiber, video 3.

fiber axis before twist insertion (which means the vector **R** is aligned). Then, we divide the chains into cylindrical shells of different diameters. Take a molecule chain straight aligned along the axis on a certain shell as an example (Fig. 9(a)). After twist insertion, we assume the same molecule chain becomes helically aligned on another shell (Fig. 9(b)). Note the length of the chain (distance between two ends) as λ, the length along the fiber axis as l, the diameter of the shell as d, and the twist number of such chain as n (it can be a fraction).

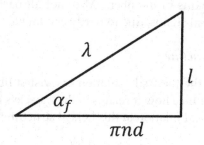

Fig. 10.   Expansion view of cylindrical shell.

No matter how the fiber is twisted, the relation among chain length λ, axial length l, shell diameter d and twist number of chain n, is given by the Pythagorean theorem, according to the expansion view of such cylindrical shell (Fig. 10), which is

$$\lambda^2 = l^2 + \pi^2 n^2 d^2 \tag{7}$$

By total derivative of Eq. (7), it shows

$$\lambda \cdot \Delta\lambda = l \cdot \Delta l + \pi^2 d^2 n \cdot \Delta n + \pi^2 n^2 d \cdot \Delta d \tag{8}$$

Given that the helically aligned chains are at an angle α_f relative to the transverse section of the fiber (Fig. 10), from $\tan \alpha_f = \frac{l}{\pi n d}$, Eq. (8) can be rearranged into

$$\frac{\Delta n}{n} = \frac{\Delta\lambda}{\lambda} \frac{1}{\cos^2\alpha_f} - \frac{\Delta d}{d} - \frac{\Delta l}{l}\tan^2\alpha_f \tag{9}$$

When the twisted fiber is heated under unchanged tension and torque, according to the rubber elasticity, the chain will contract due to the unchanged applied force on it, which means there will be a negative $\frac{\Delta \lambda}{\lambda}$. Experimental measurements show that the relative change in shell diameter and axial length, $\frac{\Delta d}{d}$ and $\frac{\Delta l}{l}$ is rather small, as well as the bias angle α_f.

Thus, the negative change in twist number of the chain by heat, mainly attributes to contraction of the chain. In fact, it is a statistic result by all the chains inside the fiber of different shells.

Although it can account for the actuation of the twist fiber qualitatively, the model can't be quantified because it is hard to determine the real alignment of the chains in the fiber. Also, not all parts of the chain can be considered as ideal chains, due to partly crystal blocks inside.

4.3. *Spring-Like Structure*

We have discussed the thermal actuation of twisted fiber under constant tension. Here we deduce how it relates to the muscle's contraction during heating, by topological relation in this helical structure:[5]

$$L_k = T_w + W_r \tag{10}$$

where T_w describes the fiber twist number, and W_r describes the amount of writhes, which depends only on the centre line configuration of the fiber. Both of them assume real numbers as values in three-dimensional space instead of integer values.[5] The linking number L_k is the sum of the two numbers, which is an invariant if there is no end rotation. Here gives Fig. 11 as an example, and notice there are continuous states between these two limits.

As shown in the expansion view of the muscle (Fig. 12), where l denotes fiber length, L denotes muscle length, N denotes the number of coil and D denotes muscle's mean diameter. We define α_c as the bias angle of fiber relative to the transverse section of the muscle as before. Accordingly, there is a relation:

$$\sin \alpha_c = L/l \tag{11}$$

It can be proved that,[5] in a uniform spring-like structure, the writhe number W_r, which depends only on the centre line configuration of the fiber, can be provided by the angle α_c and coil number N, $W_r = N \cdot (1 - \sin\alpha_c)$. Since there is no end rotation, according to Eq. (10), while writhe number changes, the twist number will commensurately change oppositely. Thus, applying the conclusion to the muscle with Eq. (11),

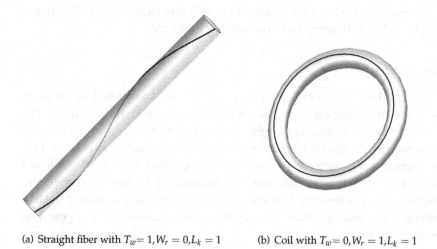

(a) Straight fiber with $T_w = 1, W_r = 0, L_k = 1$ (b) Coil with $T_w = 0, W_r = 1, L_k = 1$

Fig. 11. Examples about Twist number T_w, Writhe number W_r and Linking number L_k.

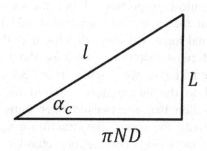

Fig. 12. Expansion view of the muscle.

then we have:

$$\Delta T_w = -\Delta W_r = N \cdot \Delta sin\alpha_c = \frac{N \cdot \Delta L}{l} \tag{12}$$

Generally, we rearrange Eq. (12) to let ΔL alone:

$$\Delta L = \frac{\Delta T_w \cdot l}{N} \tag{13}$$

Eq. (13) relates the tensile actuation of the muscle to the torsional actuation of twisted fiber. When heated, the twisted fiber will untwist, providing a negative ΔT_w. Based on Eq. (13), a negative ΔT_w will induce a negative ΔL, indicating a contraction in the muscle.

4.4. Discussion

Some models like worm-like chain model[6] are considered to be more accurate in rubber elasticity, due to their more valid assumptions. But considering the complexity of polymer in the molecular scale, they can't describe the thermal actuation of artificial muscle quantitatively. In fact, in experiments, there are several conditions that the muscle will not show reversible tensile actuation. For instance, if the load is overweight, the muscle will be stretched out instead of contraction during heated. Interactions among chains and crystal blocks inside are supposed to account for this.

5. Conclusions

In this solution, we investigate the artificial muscle from its fabrication to its thermomechanical properties. From the experiment of muscle fabrication, we find that a suitable tension should be controlled. By analysing the internal force, we show that the helical structure is stable under tension and twist insertion. Because the fishing line used is a kind of polymer material, we discuss its rubber elasticity, which is positive correlated to the temperature. Based on this property, we explain the observation that the twisted fiber untwists spontaneously during heating. Finally, the spring-like structure magnifies the torsional actuation, resulting in obvious contraction actuation, with a coefficient related to the fiber length l and coil number N.

References

1. Wikipedia. Artificial muscle — wikipedia, the free encyclopedia. URL https://en.wikipedia.org/w/index.php?title=Artificial_muscle&oldid=700615725 (2016). [Online; accessed 4-March-2016].
2. C. S. Haines, M. D. Lima, N. Li, G. M. Spinks, J. Foroughi, J. D. Madden, S. H. Kim, S. Fang, M. J. de Andrade, F. Göktepe, et al., Artificial muscles from fishing line and sewing thread, *science*. **343**(6173), 868–872 (2014).
3. Wikipedia. Rubber elasticity — wikipedia, the free encyclopedia. URL https://en.wikipedia.org/w/index.php?title=Rubber_elasticity&oldid=687490426 (2015). [Online; accessed 4-April-2016].

4. Wikipedia. Ideal chain — wikipedia, the free encyclopedia. URL https://en.wikipedia.org/w/index.php?title=Ideal_chain&oldid=684577519 (2015). [Online; accessed 4-April-2016].

5. G. Van der Heijden and J. Thompson, Helical and localised buckling in twisted rods: a unified analysis of the symmetric case, *Nonlinear Dynamics*. **21**(1), 71–99 (2000).

6. Wikipedia. Worm-like chain — wikipedia, the free encyclopedia. URL https://en.wikipedia.org/w/index.php?title=Worm-like_chain&oldid=720051853 (2016). [Online; accessed 2-August-2016].

Chapter 2

2015 Problem 4: Liquid Film Motor

Hailin Xu[1]*, Wenli Gao[2], Sihui Wang[2], Huijun Zhou[2]

[1]*Kuang Yaming Honors School, Nanjing University*
[2]*School of Physics, Nanjing University*

In this solution, we investigate the rotating of a polarized liquid film in an electric field. Ionic polarization model and Helmholtz-Smoluchowski slip condition are applied to describe the motion. We divide the film into two parts: boundary layer bearing polarization charge and inner part with no net charge. Theoretically, we solve the electric field and the charge distribution. We calculate the flow velocity near the boundary. In the experiment, by changing voltages and shape of the film, we verify the predictions for the relation between the rotating velocity and the parameters.

1. Introduction

Form a soap film on a flat frame. Put the film in an electric field parallel to the film surface and pass an electric current through the film. The film rotates in its plane. Investigate and explain the phenomenon.

*E-mail: 141242055@smail.nju.edu.com

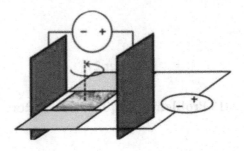

Fig. 1. The set-up of the experiment.

Different models have been put forward to solve this problem. For example, nonuniform ion distribution,[1] caused by external electric field (the electrophoresis effect), can break the symmetry under translation of the frictional forces acting on the liquid molecules. Another mechanism[2] has been suggested by Chiragwandi. Based on ionization of the water molecules in the electrolysis double-layers close to the electrodes, they explained the observed micro-scale vortices in water-based transistors. In this paper,we will improve the former models and provide a more complete solution.

The related physical mechanism is simple and clear. The field of the capacitor creates opposite electric charges in the fluid near the interface boundaries. Then the action of the electric field (related to the potential differences between the electrodes) creates the rotating flow. When the film reaches the equilibrium state, the torque caused by the field is equal to the resisting torque.

In this solution, we solve the electric potential and field over the film and the charge density near the boundary. We show that an average rotating flow in a film can be generated by edge effect.[3] We will investigate influence of different parameters such as the shape of the film, potential difference of the capacitor, potential difference between the electrodes. We will also discuss the angular velocity distribution and find how it evolves as the voltages change.

2. Preliminary Experiment

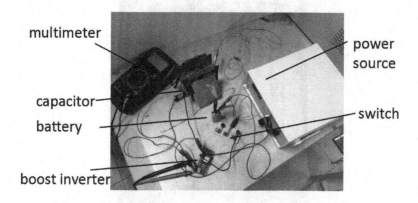

multimeter

power
source

capacitor

switch

battery

boost inverter

Fig. 2.   Experimental apparatus.

Fig. 3.   Frame inside the capacitor.

In our experiment, we use two thin aluminum plates (height: 10cm, width: 20cm) to build a capacitor. The film is held by a 2.5×3.4(cm) frame with copper wires coiled on shorter sides of the frame serving as electrodes. Potential difference of the electrodes is generated by a battery pack, while the voltage of capacitor by boost inverter (DC) with the highest voltage up to 2000V. Two multimeters are monitoring the voltages and a switch is used to control the system. (Fig. 2 and Fig. 3)

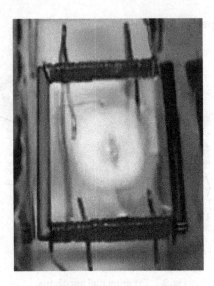

Fig. 4. Phenomenon.

The phenomenon is shown in Fig. 4. The film rotates under a proper voltage and shows a colorful pattern. It can be understood as follows. As the film rotates, there exists centrifugal force, which causes inner part of the film relatively thinner than the outer part. Since the wavelength of thin-film interference depends on film's thickness, the film shows a gradual change of color. Color varies from blue ($440nm < \lambda < 485nm$) in the central area to red ($622nm < \lambda < 760nm$) around the outer area. The average magnitude of film's thickness can be estimated to be about $1\mu m$.

3. Theoretical Analysis

As the potential differences are given as boundary conditions, by making some reasonable assumptions, the electric field and charge density can be solved completely over the film region. Then using Helmholtz-Smoluchowski slip condition we can determine the boundary conditions of velocity. Consequently, the rotating flows of the inner part is solvable by the Navier-Stokes equation.

3.1. Basic Equations

Before listing all basic equations the problem involves, we present some assumptions to make the calculation relatively simple:

(1) It is a surface phenomenon and we consider the surface lies in Oxy plane and ignore the z coordinate.
(2) The gravity and the surface tension are absent.
(3) Polarization charge distributes in a sufficiently thin layer near the boundary and there exists no net charge in the inner part.
(4) The chemical reaction on the electrodes is neglectable.
(5) The system investigated is in its static state and we won't consider the time parameter t.

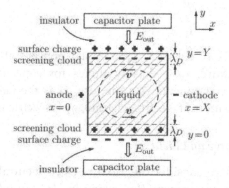

Fig. 5. Coordinates setting.[3]

To analyze hydromechanical problems, the first equation we have to encounter is Navier-Stokes equation (the *N-S equation*):

$$\rho\frac{\partial \mathbf{v}}{\partial t} + \rho\mathbf{v}\cdot\nabla\mathbf{v} = -\nabla p + \nu\rho\Delta\mathbf{v} - \frac{1}{2}\rho E^2\nabla\varepsilon + q\mathbf{E} \tag{1}$$

And the continuity equation:

$$\nabla\cdot\mathbf{v} = 0 \tag{2}$$

In which $\mathbf{v}$ is velocity, p is pressure, ρ is liquid density, ν is kinematic viscosity, q is charge density, ε is the dielectric constant, and $\mathbf{E}$ is electric field strength.

To determine the potential and charge distribution in the liquid film, the following equations are introduced.

The Poisson-Boltzmann equation:

$$\nabla \cdot (\varepsilon \mathbf{E}) = q \tag{3}$$

The relation between the field and potential:

$$\mathbf{E} = -\nabla \varphi \tag{4}$$

Equations about multicomponent medium and mass transferring (the Nernst-Plank equations without source terms) are

$$\rho_e = F \sum_k e_k c_k \tag{5}$$

$$\frac{\partial c_k}{\partial t} + \mathbf{v} \cdot \nabla c_k + \nabla \cdot \mathbf{i}_k = 0 \tag{6}$$

$$\mathbf{i}_k = -D_k \nabla c_k + e_k \gamma_k c_k \mathbf{E} \tag{7}$$

In which F is Faraday constant, c_k is molar concentration for the kth component of mixture, $\mathbf{i}_k$ is density fluxes for concentrations, D_k is diffusivity for the components of a mixture, e_k is the electric charge of components (in the units of electron charge), γ_k is electric mobility.

3.2. *Boundary Charge and Field*

For simplification, we consider the mixture electroneutral and that only two kinds of ions are present (for example, H^+ and OH^-) and the equilibrium Boltzmann distribution is valid:

$$c_1 = c_2 = c, z_1 = 1, z_2 = -1 \tag{8}$$

What we pay attention to is the charge density in the vicinity of the boundary of $y = 0$ and $y = Y$. Without loss of generality, take the $y = 0$ boundary. For any x, set the concentration $c_B = c_B(y)$. Assuming the transverse electric field strength E, write the potential as $\varphi = \phi(y) + Ex$. Making use of the no-leak boundary condition $\mathbf{i}_B \cdot \mathbf{n} = 0$ in which $\mathbf{i}_B$ is given by Eq. (7), we find:

$$c(y) = c_B e^{-e\gamma\phi(y)}, q(y) = e c_B e^{-e\gamma\phi(y)} \tag{9}$$

To solve the potential over the film, use Poisson-Boltzmann equation:

$$\varepsilon \triangle_0 \varphi = -q, \quad q = ec \tag{10}$$

along with boundary conditions:

$$\frac{\partial \varphi}{\partial \mathbf{n}} = \pm E_0, y = 0, Y \tag{11}$$

$$\varphi = 0, x = 0; \varphi = \varphi_0, x = X \tag{12}$$

In which E_0 is the electric field strength caused by the capacitor which is assumed to be a constant along the boundary $y = 0, Y$. Still we have $|E_0| = (\varepsilon_{out}/\varepsilon)|E_{out}| \cos \alpha$. By solving Eq. (10), the electric field strength of the boundary of $y = 0, Y$ is:

$$E_x \mid_{y=0,Y} = \frac{\varphi_0}{X} \pm E_0 G(x; X, Y)(y = 0, +; y = Y, -) \tag{13}$$

where G is the Green Function of the system:

$$G(x; X, Y) = \frac{4}{\pi} \Sigma_{k=0}^{\infty} \frac{\tanh \frac{(2k+1)\pi Y}{2X}}{2k+1} \cos(2k+1)\frac{\pi x}{X} \tag{14}$$

The graph of E_x at $y = 0$ when $X = Y$ is shown in Fig. 6:

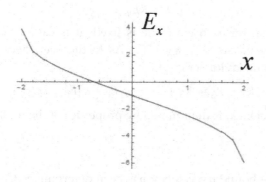

Fig. 6.   The E_x at $y = 0$ while $X = Y$.

And the corresponding potential distribution is shown in Fig. 7:

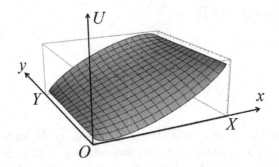

Fig. 7.   Potential over the film while X=Y.

3.3. *Boundary Velocity*[3]

So far, the electric field has been solved. Then we will find the relation between the electric field and boundary velocity.

Let the velocity be $\mathbf{u} = [u, v]$. The no-leak conditions are

$$u \mid_{x=0,X} = 0, v \mid_{y=0,Y} = 0 \tag{15}$$

It is a trivial conclusion. What we concern is the tangent velocity. First, set τ to be the unit tangent vector. The *Helmholtz-Smoluchowski slip condition* says that the tangent velocity is in direct proportion to the tangent electric field strength. So we obtain another two boundary conditions:

$$\begin{cases} u = RE_x \mid_{y=0,Y} \\ v = RE_y \mid_{x=0,X} \end{cases} \tag{16}$$

As the potential on each electrode is fixed, it is easy to conclude that $E_y \mid_{x=0,X} = 0$ and so $v \mid_{x=0,X} = 0$. As to the coefficient R, it can be discomposed into taylor series of E_0:

$$R = R_1 + R_3 + R_5 + \ldots = k_1 E_0 + k_3 E_0^3 + k_5 E_0^5 + \ldots \tag{17}$$

And the coefficient k_i is determined by properties of the mixture.

3.4. *Summery*

Up to now, the boundary velocity has been determined. Combined with the electric neutrality of the inner part, we can give a numerical solution to the liquid film motor by the N-S equation (Eq. (1)).

Next, we use the conclusions derived above to explain the phenomenon in the experiments.

4. Experiments and Comparison with Theory

In following experiments we use the same facility described in Sec. 2. Besides, we use a high speed camera to record the motion. With the help of *Tracker*, we specify the velocity and angular velocity with respect to the centre of the film of tracking points. Hereinafter we define U_1 as the voltage of the capacitor and U_2 as the voltage between the electrodes. And we switch the original point of the coordinates to the center of the rectangular film. To investigate the relation between the speed of rotation and U_1, U_2 quantitatively, measure the angular velocity ω_i of k points located equidistantly from the centre to the boundary on the x-coordinate (x_i) or y-coordinate (y_i), $i = 1, 2, \ldots, k$, as illustrated in Fig. 8. It also tells us the angular velocity distribution.

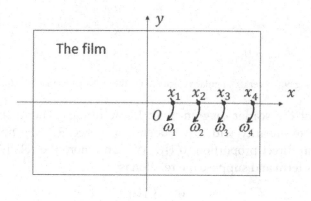

Fig. 8.   The coordinate system.

To describe the global movement feature, we define an *average angular function*, indicated as ω_x and ω_y. It is the weighted average angular velocity with respect to each ω_i's dominant region:$(x_{i+1} - x_{i-1})$ with predefined $x_0 = 0$ and $x_{k+1} = \frac{X}{2}$. The condition of y is similar.

4.1. *Direction of Rotation*

At a rough estimate, the polarization vector **P** shares the same direction of $\mathbf{E}_{out}$ and the electric dipole torque is $\mathbf{P} \times \mathbf{E}$, which is also the rotating direction. It is proved by experiment. There is a detail to be explained: why no net charge near the electrodes? My explanation is that as the system is a part in a circuit, there does not exist net charge but electric current in it.

4.2. *The Relation Between ω and U_1*

Let it be *experiment 1*. U_2 is fixed as 105V. Increasing U_1 gradually from 400V to 800V, the $\omega - U_1$ graph obtained is shown in Fig. 9.

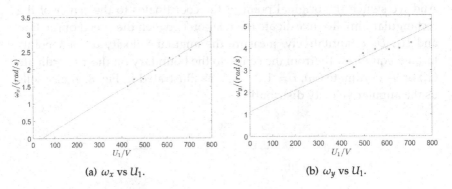

(a) ω_x vs U_1.   (b) ω_y vs U_1.

Fig. 9. Average angular velocity ω vs. the voltage of capacitor U_1.

We plot the scatter diagram with linear fitting. The ω submits an approximate linearity with U_1. As shown in Sec. 3.3, the polarization vector is in direct proportion to U_1. We can ignore the relatively small zero-order term and suppose the relation is:

$$\omega = C_1 U_1 \tag{18}$$

4.3. *The Relation Between ω and U_2*

Let it be *experiment 2*. Set U_1 fixed at 756($\pm$2)V. Raising U_2, which is the voltage between electrodes, from 47V to 324V gradually, we get the $\omega_{x_i} - U_2$ ($i = 1, 2, 3$) graph in Fig. 10.

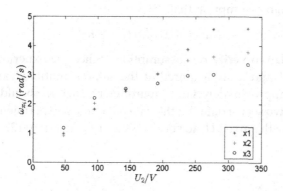

Fig. 10. Angular velocity ω_{x_i} at point $(x_i, 0)$ vs. electrodes' voltage U_2, $i = 1, 2, 3$.

As the graph shows, each curve does not appear well linearity. Use *average angular function* to combine the three curves into one and fit the curve (Fig. 11).

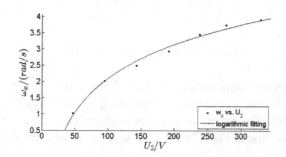

Fig. 11. Average angular velocity ω_x vs. electrodes' voltage U_2.

The logarithmic fitting curve fits the original points quiet well. We suppose the relation between ω and U_2 can be expressed by:

$$\omega = C_2(ln(U_2) + C_3) \tag{19}$$

We will make some further discussions on the numerical result.

4.4. *More Data Analysis*

Combining Eq. (18) and Eq. (19) together, we can get the influence of U_1 and U_2 on the angular velocity of the film.

We make an assumption that:

$$\omega = K_1 U_1 [ln(U_2) + K_2] \tag{20}$$

We use Matlab to verify our assumption. Since the experiment data of experiment 1 and 2 are acquired at the same condition: same mixture system and approximately same environment, each K_i should remain the same in the two sets of data. Fit the data obtained in experiment 2 in which U_2 changes with a fixed U_1 to specify K_1 and K_2 first (Fig. 12).

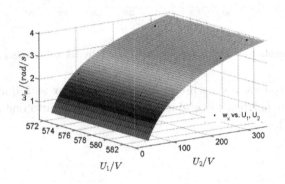

Fig. 12.   Fitting surface of ω vs. U_1, U_2.

Then we get the two coefficients, $K_1 = 2.583 \times 10^{-3}$ and $K_2 = -3.201$, with 95% confidence bounds. Substituting U_1 and U_2 of data from experiment 1 into the result, the corresponding relation between ω and U_1 can be got (blue line in Fig. 13), which is consistent with the experimental data (red dot).

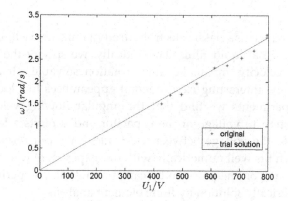

Fig. 13.   Comparison between the trial solution and original ω.

Since U_2 is strictly held fixed in experiment 1, we can choose U_1 as the index. As it can be seen in Fig. 13, the coefficients K_1 and K_2 fit the data well. So the assumption of Eq. (20) is met. Stricter proof or falsification awaits more experiments and numerical solving of Navier-Stokes equation (Eq. (1)).

4.5. *Local Reverse Flow*

In our experiment, we find there exists local reverse flow at diagonal corners (see Fig. 14).

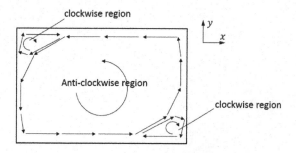

Fig. 14.   Local reverse flow.

The reason of this appearance is shown by Eq. (16), Eq. (13) and Fig. 6. There always is an interval where E_x is reversed. Because of Eq. (16), the boundary velocity is reversed and it causes reverse-flow region.

5. Conclusion

In this solution, we use classic electrohydrodynamic theory to investigate rotating flows in a liquid film. Theoretically, we specify the boundary condition of velocity and make inner rotation solvable. It also helps us explain some interesting experimental appearances like local reverse flow. In experiments we find that the angular flowing velocity is in direct proportion to voltage of the capacitor and submits a logarithmic relationship to the voltage between electrodes. We put forward a trial solution which fits well numerically with our experiment results. We will do more research on the problem by carrying out more experiments and seeking numerically solution by finite element analysis.

References

1. A. Amjadi, R. Shirsavar, N. H. Radja, and M. R. Ejtehadi, A liquid film motor, *Microfluidics & Nanofluidics*. **6**(5), 711–715 (2008).
2. Z. G. Chiragwandi, O. Nur, M. Willander, and I. Panas, Vortex rings in pure water under static external electric field, *Applied Physics Letters*. **87**(87), 153109–153111 (2005).
3. E. V. Shiryaeva, V. A. Vladimirov, and Z. M Yu, Theory of rotating electrohydrodynamic flows in a liquid film., *Physical Review E Statistical Nonlinear & Soft Matter Physics*. **80**(4), 593–598 (2009).

Chapter 3

2015 Problem 5: Two Balloons

Heyang Long[1]*, Luyan Yu[2]†, Ruyan Sun[2], Sihui Wang[1], Huijun Zhou[1]

[1]*School of Physics, Nanjing University*
[2]*Kuang Yaming Honors School, Nanjing University*

In this work, we observed the different directions of air flow when two balloons are collected through preliminary experiments. We use phenomenological theory to describe the relation between pressure and elongation of balloons. A phase diagram is made according to the pressure and elongation relation from which the direction of air flow can be predicted. In experiment, we measured the pressure difference of balloons and plotted the pressure-elongation diagram to verify the theory. Besides, we take other factors that may influence the direction of air flow into account, like inflating history, Mullins effect, rubber material. Finally, we discuss the stability of equilibrium between two balloons.

1. Introduction

Problem Statement:

Two rubber balloons are partially inflated with air and connected together by a hose with a valve. It is found that depending on initial balloon volumes, the air can flow in different directions. Investigate this phenomenon.

In this article, preliminary experiments are done at first to reproduce the phenomena and investigate the relation between balloon volumes and directions of air flow. The air flow direction depends on the pressure difference between the two balloons. So we use phenomenological theory based on the works of Mooney, Rivlin and Hart Smith[1] to describe the relation between pressure and elongation of balloons. A phase

---

*E-mail: 141120065@smail.nju.edu.com
†E-mail: 131242066@smail.nju.edu.com

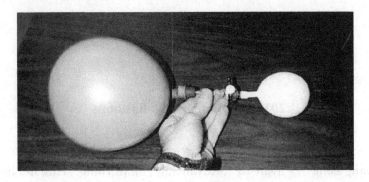

Fig. 1.   Air can flow in different directions when two balloons are connected.

diagram relating the elongations of both balloons and the flow direction is plotted. Systematic experiments were done to verify our theory. Comparisons show that the experimental data are consistent with our theoretical predictions.

We will also discuss some other factors that affect the pressure inside a balloon. Inflation history, Mullins effect[2] and material factors will be taken into account to predict how they affect the directions of air flow. As the pressure equation includes material constants, which will change with the amounts of extensions, uniaxial extension experiments of rubber, inflating and deflating balloon experiments are done to verify these properties. Hysteresis loop of balloon's pressure will be observed. Finally, we will also analyze the stability of two balloons' equilibrium using the predictions we have made.

Before theoretical analysis, preliminary experiments were done to show that air could flow in different directions[‡].

2. Theoretical Analysis

The air flow direction is determined by the pressure differences between two balloons. From pre-experiments, we suggest that pressure is not monotonously dependent on balloon's volume, we need to explore the exact relationship between them. In our research, we use elongation to describe the changes of volume, as volume is approximately proportional to cubic of elongation.

[‡]See supplementary materials, Two Balloons, video 1, 2, 3.

2.1. *Phenomenological Theory*

We apply phenomenological theory of rubber which is concise and feasible for large strain. The two basic assumptions of the theory are that rubber is incompressible and is isotropic in unstrained state. The value of poissons ratio of rubber[3] (about 0.499) indicates that rubber is nearly incompressible. The condition for isotropy requires that the strain-energy function W be symmetrical with respect to the three principal extension ratios λ_1, λ_2, λ_3, we can write the three simplest possible even-powered functions which satisfy these requirements.[1] I_1, I_2, I_3 in the functions are called strain invariants.

$$I_1 = \lambda_1^2 + \lambda_2^2 + \lambda_3^2$$
$$I_2 = \lambda_1^2\lambda_2^2 + \lambda_2^2\lambda_3^2 + \lambda_3^2\lambda_1^2 \tag{1}$$
$$I_3 = \lambda_1^2\lambda_2^2\lambda_3^2$$

Because the deformation of a spherical balloon is an equi-biaxial tension, the first two principle elongations are the same.

$$\lambda_1 = \lambda_2 = \lambda \tag{2}$$

The condition for incompressibility or constant volume during deformation can be expressed by the relation,

$$I_3 = 1 \tag{3}$$

The combination of Eq. (1) to Eq. (3) yields:

$$I_1 = 2\lambda^2 + \frac{1}{\lambda^4}$$
$$I_2 = \lambda^4 + \frac{2}{\lambda^2} \tag{4}$$

Rivlin derives the expression of the principal stresses t_i,[1] which involves the partial derivatives of the strain-energy function with respect to the independent variables I_1, I_2.

$$t_i = 2(\lambda_i^2 \frac{\partial W}{\partial I_1} + \lambda_i^{-2} \frac{\partial W}{\partial I_2}) + P(i = 1, 2, 3) \tag{5}$$

where P is an arbitrary hydrostatic stress.[1]

By subtraction, P may be eliminated to give the three principal stress differences.

$$t_1 - t_2 = 2(\lambda_1{}^2 - \lambda_2{}^2)(\frac{\partial W}{\partial I_1} + \lambda_3{}^2 \frac{\partial W}{\partial I_2})$$

$$t_2 - t_3 = 2(\lambda_2{}^2 - \lambda_3{}^2)(\frac{\partial W}{\partial I_1} + \lambda_1{}^2 \frac{\partial W}{\partial I_2}) \qquad (6)$$

$$t_3 - t_1 = 2(\lambda_3{}^2 - \lambda_1{}^2)(\frac{\partial W}{\partial I_1} + \lambda_2{}^2 \frac{\partial W}{\partial I_2})$$

On the surface of a balloon, $t_3 = 0$, $t_1 = t_2$. Substitute these relations into Eq. (6), it yields:

$$t = t_1 = t_2 = 2(\lambda^2 - \frac{1}{\lambda^4})(\frac{\partial W}{\partial I_1} + \lambda^2 \frac{\partial W}{\partial I_2}) \qquad (7)$$

Given non-Gaussian theory, Hart Smith[1] suggested three-constant formula to explain a variety of data on the inflation of balloons.

$$\frac{\partial W}{\partial I_1} = Ge^{k_1(I_1-3)^2} \qquad (8)$$

$$\frac{\partial W}{\partial I_2} = G\frac{k_2}{I_2} \qquad (9)$$

Substitute Eq. (8) and Eq. (9) to Eq. (7) we have:

$$t = 2(\lambda^2 - \frac{1}{\lambda^4})(Ge^{k_1(I_1-3)^2} + \lambda^2 G\frac{k_2}{I_2}) \qquad (10)$$

Then combine with Eq. (4), we find the expression of principle stress with respect to extension ratio λ

$$t = 2G(\lambda^2 - \frac{1}{\lambda^4})[e^{k_1(2\lambda^2+\frac{1}{\lambda^4}-3)^2} + \frac{k_2\lambda^4}{\lambda^6+2}] \qquad (11)$$

Consider force equilibrium of a half spherical balloon 2 as shown in Fig. 2, we have

$$(\pi(H + R)^2 - \pi R^2)t = \pi R^2 \Delta p \qquad (12)$$

Using $H \ll R$, we get

$$\Delta p = \frac{2H}{R}t \qquad (13)$$

Where $\Delta p = p - p_0$, p_0 is atmospheric pressure. Note that here $H = \frac{H_0}{\lambda^2}$, $R = \lambda R_0$.

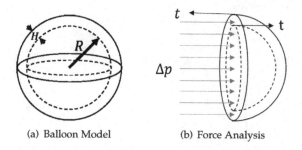

(a) Balloon Model   (b) Force Analysis

Fig. 2.   The pressure inside a balloon is proportional to the stress of rubber.

Finally Δp can be written as a function of λ.

$$\Delta p = \frac{4H_0G}{R_0}\left(\frac{1}{\lambda} - \frac{1}{\lambda^7}\right)[e^{C_1(2\lambda^2 + \frac{1}{\lambda^4} - 3)^2} + \frac{C_2\lambda^4}{\lambda^6 + 2}] \tag{14}$$

where C_1, C_2, G are material constants dependent on rubber materials, H_0 is the initial thickness of the balloon, R_0 is initial radius, λ is extension ratio used to describe elongation.

2.2. Analysis of Air Flow Directions

Based on the the pressure function with respect to elongation we have derived above, we can explain the phenomena in preliminary experiments. Fig. 3 is a theoretical curve of pressure with respect to elongation with parameters taken as $\frac{4H_0G}{R_0} = 1700Pa$, $C_1 = 0.001$, $C_2 = 0.8$. The dots represent two balloons of different initial conditions. In Fig. 3(a), the larger balloon have greater pressure so that air flows from it to the smaller balloon. In Fig. 3(b), the smaller balloon have greater pressure so that air flows in the opposite direction. In Fig. 3(c), the two balloons have the same pressure, even though their elongations are different, the air doesn't flow either way.

3. Experiment

To verify the theoretical predictions, we need to measure the pressure function with respect to elongation in experiment. An U-tube barometer is connected to the balloon to measure the pressure inside. At the same time, we used ABviewer to measure the diameters of balloons along two

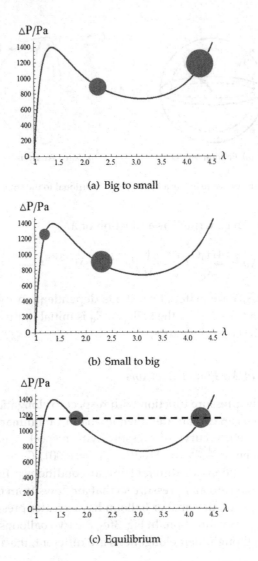

(a) Big to small

(b) Small to big

(c) Equilibrium

Fig. 3.   Theoretical curve of pressure with respect to elongation. The direction of air flow depends on pressure difference between the balloons.

directions, see Fig. 4. The average radius of the balloon is defined as $R = \frac{(D_1 D_2^2)^{\frac{1}{3}}}{2}$, then elongation is determined by $\lambda = \frac{(D_1 D_2^2)^{\frac{1}{3}}}{2R_0}$.

In Fig. 5, the dots are experimental data, the curve is a theoretical result by Eq. (14). The material constants in Eq. (14) were obtained by by fitting

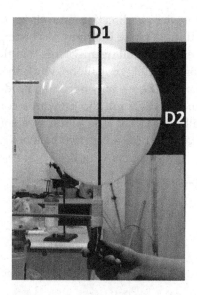

Fig. 4. Measurement of the diameters of balloons along two directions. Average radius of the balloon $R = \frac{(D_1 D_2^2)^{\frac{1}{3}}}{2}$.

the equation to the experimental data, here $G = 36645 Pa$, $C_1 = 0.0012$, $C_2 = 0.79$.

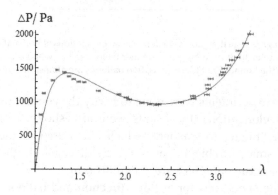

Fig. 5. Pressure-elongation(Δp-λ) curve.

The experimental dada are in good agreement with the theoretical function and support our explanation for the airflow direction. For example, in Fig. 5 the pressure reaches a maximum when the elongation is near 1.4, which explains when airflow from small balloons to big ones may occur.

The overall condition for airflow direction and elongations of both balloons can be summarized in a "phase diagram" according to Eq. (14), see Fig. 6 and compared to experimental results. Note that the horizontal axis represents the elongation of the bigger balloon, while the vertical axis represents that of the smaller one. So we only need to consider the region below the diagonal line. In this diagram, the grey area corresponds to airflow from bigger to smaller balloons. The white area corresponds to the opposite. On the boundary of these areas, two balloons are almost in equilibrium.

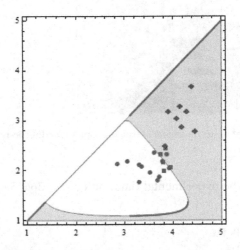

Fig. 6.   Phase Diagram: Airflow direction depends on elongations of both balloons. The grey area corresponds to airflow from bigger to smaller balloons. The white area corresponds to the opposite. On the boundary of these areas, two balloons are almost in equilibrium.

Numerous experiments were done to verify the conditions for airflow direction and elongations, the results were also shown in Fig. 6. The diamonds are "bigger to smaller", which all located in the gray area. The dots are "smaller to bigger" ones, which all located in the white area. The squares near the boundary are in equilibrium.

Till now the conditions for airflow direction and balloon elongations are confirmed.

4. Other Factors

By now, we've discussed how elongation influence the pressure inside a balloon. However, there are also many other factors that will affect the pressure. In order to predict the airflow direction more precisely, we need to take these factors into account.

4.1. *Inflation History*

When we connect two similar balloons of the same size, if one has been inflated many times while the other one is new, we will find that the used balloon get inflated further even if they both have the same initial volume[§].

To verify this property of rubber, we conducted a uniaxial extension experiment. The experimental setup is shown in Fig. 7. It includes a force sensor connected to a balloon rubber sample, both installed on a rail with scales.

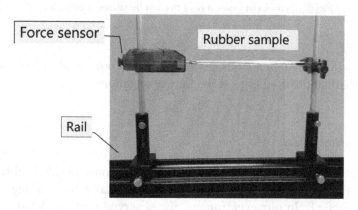

Fig. 7. Experiment setup for uniaxial extension.

Fig. 8 is the force and extension curve of the first five times of extensions for one rubber sample.

From the experimental data in Fig. 8, we find that the extension of a new rubber produces the biggest stress. Then the stress of rubber changes with extension times, then tends to become stable. We can observe that the 4th and 5th curves almost overlap. According to the force analysis of the balloon in Fig. 2, the extra pressure inside a balloon is proportional to

---

[§]See supplementary materials, Two Balloons, video 4.

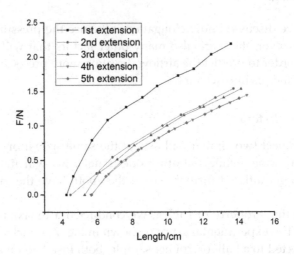

Fig. 8.   Force-extension curve of the first five times of extensions.

the stress of the rubber. Consequently, new balloons tend to have higher pressure than the used ones at the same elongations.

4.2. *Mullins Effect*

The Mullins effect describes the mechanical response in filled rubbers in which the stress-strain curve depends on the maximum loading previously applied. In our experiments, we observed that the deflating and the inflating balloon have different pressures. In our video [1], the two balloons are both stretched many times with nearly the same elongation. A deflating balloon on the left is connected to a inflating balloon on the right. Then air flows from the inflating balloon to the deflating one.

To investigate this effect, we conducted a pressure-elongation experiment during inflation and deflation. It forms a hysteresis loop, see Fig. 9. In Fig. 9, the upper curve corresponds to inflation, the lower curve corresponds to deflation.

---

[1]See supplementary materials, Two Balloons, video 5.

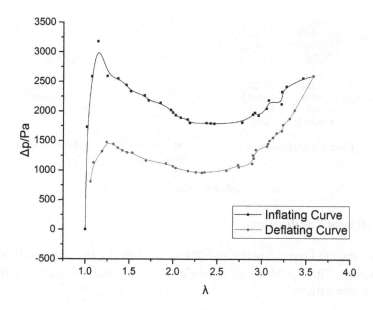

Fig. 9. Mullins effect, pressure-elongation during inflation and deflation forms a hysteresis loop.

4.3. *Balloon Material*

The third factor is rubber material of balloons. We used three kinds of balloons for comparison.

Fig. 10(b) is the pressure-elongation curves of the three balloon samples, called A, B and C. The trends are similar, but the material constants are dependent on the rubber they're made of. For small elongation, the pressure of Sample A is larger than the other two samples. So air will flow from it to the others. When balloons are inflated bigger, the pressure of Sample A becomes smaller than others, air will flow in the opposite direction.

In our experimental videos[||], two balloons made of rubber A and B are connected. Although they have similar initial elongation, the air flows from A to B. When A and C are connected, air flows from A to C due to higher pressure level.

[||] See supplementary materials, Two Balloons, video 6.

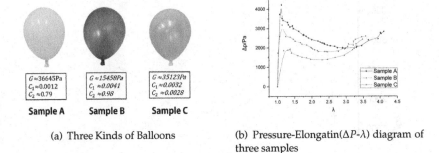

<table>
</table>

(a) Three Kinds of Balloons

(b) Pressure-Elongatin(ΔP-λ) diagram of three samples

Fig. 10.   Different Balloon Materials.

5. Stability of Equilibrium

Now we start to investigate the stability of equilibrium in the two-balloon experiments. The equilibrium can be classified as similar elongations and different elongations.

5.1. *Similar Elongations*

If the balloon elongations are the same, there are three different situations according to initial elongations. In the first situation, see Fig. 11, if one balloon is given a small perturbation and becomes bigger, it has greater pressure than the other, so that and air flows from it to inflate the smaller one. In this case, the equilibrium is stable.

In the second situation, see Fig. 12, give one balloon a perturbation, the balloon that becomes bigger will have smaller pressure and air flows from the smaller balloon to the big one. In this case, the equilibrium is unstable. Similar to the first situation, we can easily find that equilibrium is stable in the third situation, see Fig. 13,

5.2. *Different Elongations*

Mullins effect needs to be taken into account in order to investigate the equilibrium stability of two balloons with different elongations. For two balloons of different elongations, there are also three situations. It is more complicated when taking two ways of perturbation into account.

When we connect two balloons at equilibrium, the total air volume is

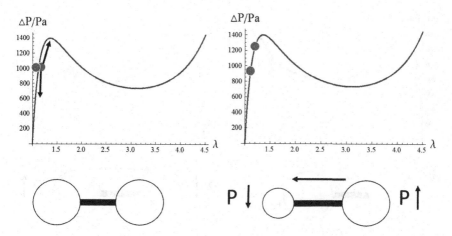

Fig. 11. Stable Equilibrium. If one balloon becomes slightly bigger it has greater pressure, so that air flows from it to the smaller one.

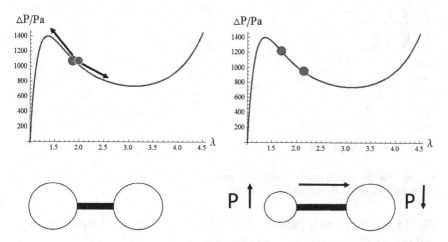

Fig. 12. Unstable Equilibrium. If one balloon becomes slightly bigger it has smaller pressure, so that air flows from the smaller one to it.

nearly constant. The volume of one balloon is approximately proportional to cubic of its elongation. When the small one got slightly inflated, the big balloons elongation change is less than the other. Besides, their pressure both increases, though, because of Mullins effect, the pressure of the deflating balloon will be less than the curve shows. Therefore, the big balloons pressure will be less than the smaller ones after perturbation. So the small balloon will inflate the big one and return to equilibrium. In

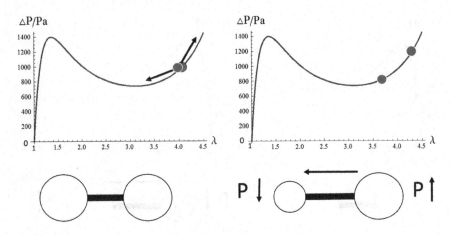

Fig. 13.   Stable Equilibrium.

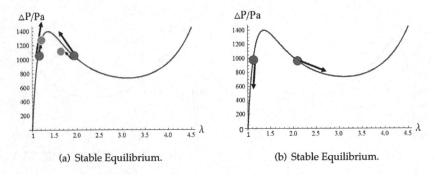

(a) Stable Equilibrium.   (b) Stable Equilibrium.

Fig. 14.   Equilibrium is stable in Fig. 14(a) and in Fig. 14(b). Generally, the equilibrium is stable for balloons with different elongations.

Fig. 14, equilibrium is stable. Generally, we can find that equilibrium is stable for balloons with different elongations.

6. Conclusion

The direction of air flow is determined by pressure difference between the balloons. We use elongation as a parameter governing the pressure inside a balloon. To investigate the relation between pressure and elongation, Mooney-Rivlin model is used. Based on our theory, pressure-elongation curve and phase diagram with respect to elongation are plotted and verified experimentally.

Then, we take other factors such as Mullins effect, inflating history and material parameters into account to predict the direction of air flow. The inside pressure in balloons of similar material is determined by its elongation. Pressure is also influenced by whether the balloon is inflating or deflating, and the biggest elongation of the present extension. Inflating history also affect the pressure. For instance, the pressure in a balloon that has been extended many times is smaller than that of a new one. Balloons of different materials have different material constants, so they have different pressure in same extension ratio. Finally, several situations concerning the stability of balance are discussed.

References

1. L. R. G. Treloar, *The physics of rubber elasticity*. Oxford University Press, USA (1975).
2. J. Diani, B. Fayolle, and P. Gilormini, A review on the mullins effect, *European Polymer Journal*. **45**(3), 601–612 (2009).
3. P. H. Mott and C. M. Roland, Limits to poisson's ratio in isotropic materials, *Physical Review B Condensed Matter*. **80**(80), 132104 (2009).
4. M. A. Johnson and M. F. Beatiy, The mullins effect in equibiaxial extension and its influence on the inflation of a balloon, *Int.J.Engng Sci*. **33**(2), 223–245 (1995).
5. D. R. Merritt and F. Weinhaus, The pressure curve for a rubber balloon, *American Journal of Physics*. **46**(10) (October, 1978).
6. C. S. Chen, Two interconnected rubber balloons as a demonstration showing the effect of surface tension (2009).
7. Y. Levin and S. F. Da, Two rubber balloons: phase diagram of air transfer., *Physical Review E Statistical Nonlinear and Soft Matter Physics*. **69**(5 Pt 1), 343–358 (2004).

Chapter 4

2015 Problem 6: Magnus Glider

Boyuan Tao[1]*, Dachuan Lu[1]†, Yiran Deng[2], Sihui Wang[3], Huijun Zhou[3]

[1]*Kuang Yaming Honors School, Nanjing University*
[2]*School of Earth Science and Engineering, Nanjing University*
[3]*School of Physics, Nanjing University*

We solve the problem of Magnus glider by experimental and theoretical investigations. In experiment, initial conditions are controlled systematically. The gliders motion can be classified into flying stage and terminal stage. The flying stage has three kinds of typical trajectories depending on initial conditions. The equation of motion for the glider has been derived and solved. All kinds of trajectories found in both stages can be given and compared to experimental conditions. By numerical calculation, we also obtain a phase diagram dividing the conditions for three kinds of typical trajectories. Finally we analyze its stability during the flight.

1. Introduction

Glue the bottoms of two light cups together to make a glider. Wind an elastic band around the centre and hold the free end that remains. While holding the glider, stretch the free end of the elastic band and then release the glider. Investigate its motion.

Magnus effect is common in ball games when a spinning ball curves away from its parabolic path. It is produced by Magnus force caused by the pressure differences around an object due to its rotation. It provides a key factor, the lift force, in deciding the fascinating trajectories of a Magnus glider in this problem.[1]

In this solution, we first made experimental investigations. Initial conditions like initial velocity and angular velocity are controlled systematically. The Magnus glider's trajectories will be recorded and

---

*E-mail: 15905172626@163.com
†E-mail: dclu@smail.nju.edu.com

analyzed. According to our observations, we classify the gliders motion into two stages: flying stage and terminal stage. Three kinds of typical trajectories will be found in the flying stage. Then we build a theoretical model to explain our observations. We will present a derivation of Magnus force, and build the equation of motion for the glider. All kinds of trajectories found in both stages will be given and compared to experimental conditions. Finally we analyze whether and how a glider can keep stability during flight.

2. Experiment

2.1. *Apparatus*

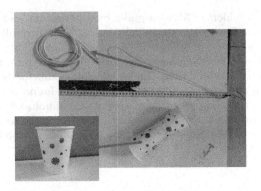

Fig. 1.   Materials to make a Magnus glider and control initial releasing conditions.

We glue two light cups together with adhesive tape to make a glider. An elastic band is wound around the glider, see Fig. 1. We fix the free end of the elastic band to a measuring tape. The initial velocity u is controlled by changing the elongation of the elastic band. The initial angular velocity w is controlled by changing the number of rounds we wind the band around the glider, see Fig. 2.

We use a high-speed camera to record videos and use software Tracker to collect data to plot the gliders trajectories, as shown in Fig. 3.

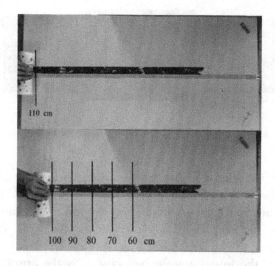

Fig. 2. The initial velocity u is controlled by changing the elongation of the elastic band. The initial angular velocity is controlled by the number of rounds we wind the band.

Fig. 3. A glider's trajectory.

2.2. Experimental Results

According to our observations, the glider's motion can be classified into two stages: flying stage (motion in the beginning) and terminal stage in which the glider will glide stably and finally drop down. We will discuss them separately.

Besides, in the flying stage, the glider has three kinds of typical trajectories: smooth, cusp and circle-back, see Fig. 4. We will show how the trajectory is determined by initial velocity u and angular velocity ω.

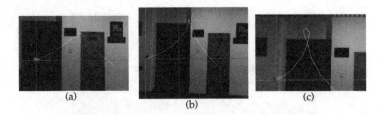

Fig. 4.   Three kinds of typical trajectories in the flying stage; (a) Smooth; (b) Cusp; (c) Circle-back.

First of all, we wind the band around the glider at the same position (110cm from the nail) to the same rounds (2 or 3), then release it at different positions, as shown in Fig. 5(a). In this way, we can change initial velocities u of the glider and keep angular velocity nearly the same. Experimental results are shown in Fig. 5(a).

In Fig. 5(a) the initial angular velocity ω is the approximately the same (2 rounds), the glider is release when the band is stretched to 70, 80, 90, 100, 110cm respectively. In Fig. 5(b) the initial angular velocity ω is also the approximately the same (3 rounds), the glider is released at 60, 70, 80, 90, 100, 110cm respectively. In Fig. 5(c), we fix u (release at 70cm) and change ω (2, 3, 4 rounds). In Fig. 5(d), we also fix u (release at 90cm) and change ω (3, 4 rounds).

As velocity v increases, the glider can reach higher and higher. For sufficiently high initial velocity, it may reach a cusp or even circling around before descending. There's a critical condition at which cusp is observed, and at that point, the glider stops momentarily. As angular velocity ω increases, the glider's trajectory also changes from smooth to cusp to circle-back. In experiment, as ω increases, the glider can reach higher. As will be given in theory (see Sec. 4.1), this is true when ω is relatively small so that the trajectory is smooth or cusp. For larger ω, the highest point of the trajectory won't get higher as ω increases.

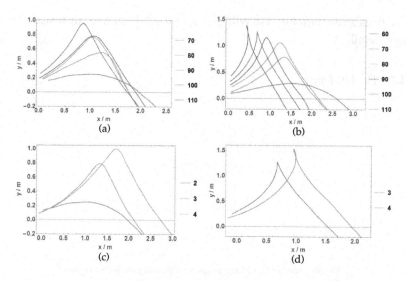

Fig. 5.   Trajectories at different initial velocities and initial angular velocities.

3. Theoretical Model

3.1. *Force Analysis*

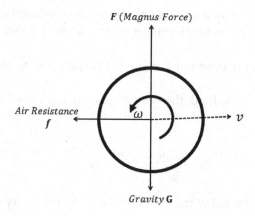

Fig. 6.   Forces experienced by a Magnus glider.

As shown in Fig. 6, the forces that a glider experiences include gravity *G*, air resistance *f* and the lift force *F* due to Magnus effect. We take the gravity as a constant mg. We will discuss the lift force and air resistance in more detail.

3.1.1. *The Lift Force*

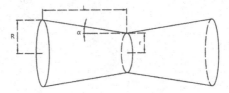

Fig. 7.   Model of a Magnus glider made of two paper cups.

Fig. 7 is the sketch of a model glider made of two paper cups glue on their bottoms. Firstly we estimate the Reynolds number,

$$Re = \frac{\rho v d}{\mu} \sim 10^4 \tag{1}$$

where the density of air ρ is $1.2 kg/m^3$; the average velocity of the glider v is taken as $5m/s$; the average diameter of the glider d is $0.05m$; the viscosity factor μ is $2 \cdot 10^{-5} Pa \cdot s$.

In this range, the potential flow model is applicable to analyze the air flow.

For incompressible fluid, the velocity potential satisfies Laplaces equation:

$$\frac{\partial^2 \Phi}{\partial x^2} + \frac{\partial^2 \Phi}{\partial y^2} = 0 \tag{2}$$

Considering the motion of the glider, we have the boundary conditions:

$$\frac{\partial \Phi}{\partial r} \big|_{r=R} = 0$$
$$\frac{\partial \Phi}{\partial r} \big|_{r \to \inf} = u \cdot \cos \theta \tag{3}$$

where u is the translational velocity of the glider.

Solving Eq. (2) and Eq. (3), we get

$$\Phi = u(1 + \frac{r^2}{R^2}) \cdot r \cdot \cos\theta + \frac{\tau}{2\pi}\theta \tag{4}$$

Here $\tau = \int_0^{2\pi} wr \cdot rd\theta = 2\pi wr^2$ is the vortex term. Differentiating it we can get the velocity at the surface:

$$v_\theta = \frac{1}{r} \cdot \frac{\partial\Phi}{\partial\theta} = -u(1 + \frac{r^2}{R^2})\sin\theta + \frac{\tau}{2\pi R} \tag{5}$$

Thus

$$v_\theta\,|_{r=R} = -2u \cdot \sin\theta + \omega R$$
$$v_r\,|_{r=R} = 0 \tag{6}$$

It means that air won't penetrate the glider radially.

Regardless of small compression and negligible viscosity of the air, we apply Bernoulli equation:

$$p + \frac{1}{2}\rho(v_\theta^2 + v_r^2) = p_0 + \frac{1}{2}\rho u^2 \tag{7}$$

Substitute Eq. (5) and Eq. (6) into it, we get

$$p = p_0 + \frac{1}{2}\rho u^2 - \frac{1}{2}\rho(-u(1 + \frac{r^2}{R^2})\sin\theta + \frac{\tau}{2\pi R})^2 \tag{8}$$

The net force produced by the pressure difference of air surrounding the glider can be estimated by dividing it into infinitesimal cylinders with thickness dx, thus

$$dF = dx \int_0^{2\pi} p \cos\alpha R \sin\theta d\theta \tag{9}$$

α is the inclination angle of the cup edge, as shown in Fig. 7.

Substitute Eq. (9) into Eq. (8), we can calculate the total lift force:

$$F = \int_{-L}^{L} dx \int_0^{2\pi} p \cos\alpha R \sin\theta d\theta = k \cdot \omega \times u$$
$$k = 4\pi\rho(\tfrac{1}{3}R^2 - \tfrac{1}{6}rR + \tfrac{6}{7}r^2)L \tag{10}$$

The total lift force can be summarized in a vector form,

$$\vec{F} = k \cdot \vec{\omega} \times \vec{u} \tag{11}$$

3.1.2. *Resistance*

For translational motion, we consider the variation of velocity of the glider is small, and take resisting force (friction) f of the air as proportional to velocity,[2] i.e.

$$f = -c \cdot v \tag{12}$$

where c is a constant which will be determined by fitting data according to experimental results.

Air is dragged with the glider as it spins, thus the gliders angular velocity has exponential decay with time:

$$\omega = \omega_0 \cdot e^{-\beta t} \tag{13}$$

For a very slow decay as observed in our experiment $(10^{-3} s^{-1})$, we assume that $\beta t \ll 1$.

3.2. *The Equation of Motion*

The ideal case is that we only consider the lift force and gravity and neglect the resistance. The equation of motion is described by

$$\begin{aligned} m\ddot{y} &= -mg + k\omega\dot{x} \\ m\ddot{x} &= -k\omega\dot{y} \end{aligned} \tag{14}$$

The solution is a cycloid trajectory given by

$$\begin{aligned} x &= \frac{mg}{k\omega}t + \frac{m(v_0 - \frac{mg}{k\omega})}{k\omega}\sin\left(\frac{k\omega}{m}t\right) \\ y &= \frac{m(v_0 - \frac{mg}{k\omega})}{k\omega}\left(1 - \cos\left(\frac{k\omega}{m}t\right)\right) \end{aligned} \tag{15}$$

Fig. 8(a) is the trajectory for $v_0 = 8m/s, \omega = 12\pi/s$. Fig. 8(b) is the trajectory for $v_0 = 5m/s, \omega = 12\pi/s$.

Unlike our experimental observations, in the solutions without air resistance, zeniths appear periodically. To describe the realistic trajectory, air resistance has to be taken into account.

Then considering all the forces above, we have the equations of motion of the glider:

$$\begin{aligned} F_x &= m\ddot{x} = -c\dot{x} - k\omega_0 e^{-\beta t}\dot{y} \\ F_y &= m\ddot{y} = -mg - c\dot{y} + k\omega_0 e^{-\beta t}\dot{x} \end{aligned} \tag{16}$$

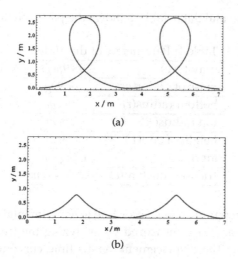

(a)

(b)

Fig. 8.   Theoretical trajectory without air resistance (cycloid).

In the flying stage, $\beta t \sim 10^{-3} << 1$, decay of angular velocity is negligible. For simplicity, we neglect decay in angular velocity so we get Eq.(17).

$$m\ddot{x} = -c\dot{x} - kw_0\dot{y}$$
$$m\ddot{y} = -mg - c\dot{y} + kw_0\dot{x}$$

(17)

The analytical solution for these linear equations can be expressed as:

$$
\begin{aligned}
x &= -\frac{2gk_1k_2}{(k_1^2+k_2^2)^2} + \frac{gk_1}{k_1^2+k_2^2}t + \frac{ige^{-(ik_1-k_2)t}}{2(k_1^2+k_2^2)^2} - \frac{ige^{-(ik_1+k_2)t}}{2(k_1^2-k_2^2)^2} + \frac{v}{k_1}e^{-k_2t}\sin k_1 t \\
&\quad + \frac{k_2v}{k_1^2+k_2^2} - \frac{k_2v}{2k_1(k_1+ik_2)}e^{(ik_1-k_2)t} - \frac{k_2v}{2k_1(k_1-ik_2)}e^{-(ik_1+k_2)t} \\
y &= -\frac{gk_2t}{k_1^2+k_2^2} + \frac{2gk_2^2}{(k_1^2+k_2^2)^2} - \frac{igk_2e^{(ik_1-k_2)t}}{2k_1(k_1+ik_2)^2} \\
&\quad + \frac{igk_2e^{-(ik_1+k_2)t}}{2k_1(k_1-ik_2)^2} + \frac{k_1v-g}{k_1^2+k_2^2} - \frac{k_1v-g}{2k_1(k_1+ik_2)}e^{(ik_1-k_2)t}
\end{aligned}
$$

(18)

where, $k_1 = \frac{kw_0}{m}, k_2 = \frac{c}{m}$

4.   Trajectories with Comparison to Experimental Results

4.1.   Flying Stage

Now we take all the parameters according to our experimental conditions. Substitute them into analytical solution Eq. (18), we can make comparison with our experimental data. Table 1 shows the parameters:

Table 1: Parameters of the glider.

gravity(g)	$9.79m/s$
mass of glider(m)	$11.93g$
bottom radius(r)	$2.6cm$
top radius(R)	$3.5cm$
height of cup(L)	$8.4cm$
air density(ρ)	$1.2kg/m^3$
friction constant(c)	$0.011Ns/m$

All the parameters are taken according to experimental conditions. The coefficient c in Eq. (12) is measured by analyzing the free fall motion of the glider. Fitting the displacement versus time curve to the following equation, we obtain the value of coefficient $c = 0.011Ns/m$.

$$y = -\frac{mgt}{c} + \frac{m^2g}{c^2}(e^{-\frac{ct}{m}} - 1) + y_0 \qquad (19)$$

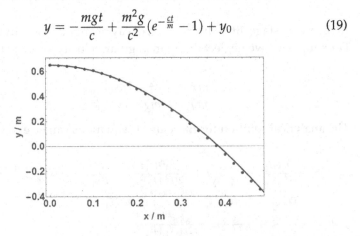

Fig. 9.   Displacement versus time curve of free fall motion of the glider.

The trajectory for $\omega_0 = 12.2\pi s^{-1}$ and $v = 5.8m/s$ are given in Fig. 10, where the experimental data (dots) are also shown for comparison with theoretical our solutions. We can see the solution fits well with experimental results.

We may solve the equations of motion under different initial velocities v and angular velocities ω_0. Fig. 5 are experimental results. Fig. 11 shows how trajectories change with v in theory.

In theory, we also obtain three types of trajectories: smooth, cusp and

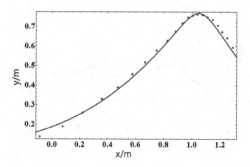

Fig. 10. Trajectory for $\omega_0 = 12.2\pi s^{-1}$ and $v = 5.8 m/s$. The curve is the theoretical result; the dots are experimental data.

circle-back. The larger the initial velocity, the higher the glider will fly, see Fig. 11 (a). They are the same as we observed in experiment, see Fig. 5.

As angular velocity ω increases, the trajectory of the glider also change from smooth to cusp to circle-back. When the trajectory is smooth, the increase of ω can make the glider reach higher and circle backward to a greater extent. But for circle-back trajectory, when ω is large enough, the highest point of the trajectory won't change as ω.

By numerical calculation, we can find the conditions for the three types of trajectories, and draw a phase diagram in terms of initial velocities v and angular velocities and ω_0, as shown in Fig. 13.

4.2. *Gliding at the Terminal Stage*

After flying for a period of time, the glider enters a terminal stage. Mathematically, the solution in Eq. (18) can be separated into a uniform motion and a damped oscillation. When the oscillation term gradually vanishes, the remaining term dominates, then the glider experiences a uniform motion, which might be called gliding, at the terminal stage. Fig. 14 shows a long term trajectory of a glider. The terminal stage is a uniform motion, which might be called gliding.

However, the solution of flying stage contains an assumption neglecting the decay in angular velocity. This approximation is reasonable when consider the small time scale at the beginning stage. Here, for a larger time scale, we must take it in account. Sure enough that the observed trajectory will be lower than that shown in Fig. 14. When we take the decay of ω into consideration, the asymptotic motion can be solved by setting the resulting

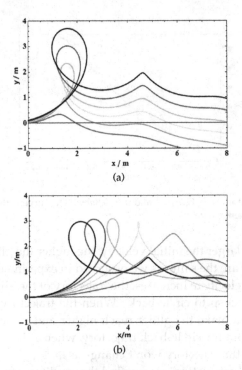

Fig. 11.   (a) Trajectories under different initial velocities $v(4 \sim 9m/s$ from bottom to top); (b) Trajectories under different angular velocities $\omega(6\pi \sim 16\pi m/s$ from bottom to top).

force to be zero in Eq. (16). Hence we have velocities:

$$
\begin{aligned}
v_x &= \frac{mgk\omega_0 e^{-\beta t}}{k^2\omega_0^2 e^{-2\beta t}+c^2} \\
v_y &= \frac{mgc}{k^2\omega_0^2 e^{-2\beta t}+c^2}
\end{aligned}
\tag{20}
$$

From Eq. (20) we get the terminal stage trajectory, as shown in Eq. (21) and Fig. 15.

$$
y = \frac{gm \log(\cos(\frac{c\beta(x+x_0)}{gm}))}{2c\beta} + y_0
\tag{21}
$$

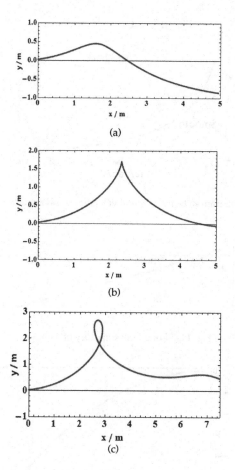

Fig. 12.   Three types of trajectories in theory: smooth, cusp and circle-back.

5. Discussion and Conclusion

5.1. *Stability of the Gliders Motion (Precession)*

When we release a glider, it may be subject to disturbances that cause uneven projection; it may also be disturbed during its motion by irregular air flow like wind or convection.   Nevertheless, as observed in our experiment, it will finally adjust itself to a stable motion. How and why it can resist such disturbances and retains a stable motion deserves further analyses.

We suppose the disturbance as lateral and represent it by an additional

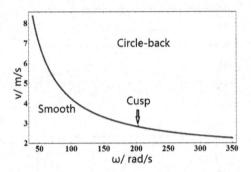

Fig. 13.   Phase diagram in terms of initial velocities v and angular velocities and ω_0.

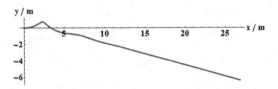

Fig. 14.   Long term trajectory of a glider.

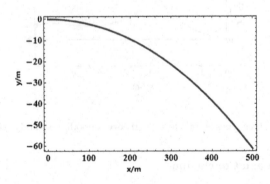

Fig. 15.   Terminal stage trajectory, considering the decay in angular velocity.

angular velocity ω' perpendicular to the axis of the glider, as shown in Fig. 16.

If we analyze the left and right sides of the glider separately, they have additional velocity u' produced by the rotation ω'. u' has similar effect as discussed in section 3.1.1:

$$\vec{F} = k \cdot \vec{\omega} \times \vec{u}' \tag{22}$$

The forces on both sides cancel, but the total torque is proportional to the additional angular velocity $\vec{M} \propto \omega'$, because u' is proportional to ω', see Fig. 16.

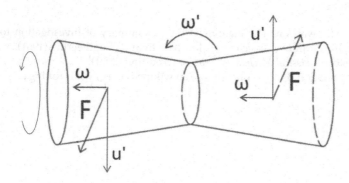

Fig. 16. A lateral disturbance is represented by an additional angular velocity ω' perpendicular to the axis of the glider. The additional velocities u' induces a pair of forces on both sides of the glider.

This torque causes a precession similar to that of a spinning bullet stabilizing its projectile gyroscopically.[3] Taking air friction into account, the precession will gradually decay, and the glider will return to stead motion.

5.2. Conclusion

We solve the problem of Magnus glider by experimental and theoretical investigations. We build a simple apparatus to control initial releasing conditions. The glider's motion can be classified into flying stage and terminal stage. The flying stage has three kinds of typical trajectories, smooth, cusp, circle-back, depending on initial velocity and angular velocity.

We build a theoretical model to explain our observations. We ascribe the Magnus force as a key factor in deciding the fascinating trajectories of the Magnus glider. The equation of motion for the glider has been derived and solved. All kinds of trajectories found in both stages can be given and compare to experimental conditions. By numerical calculation, we also obtain a phase diagram in terms of initial velocities and angular velocities.

Finally we found that the glider stabilize itself by a gyroscopic precession during the flight.

References

1. B. W. M. Swanson. The magnus effect: a summary of investigation to date; trans. In *ASME D. J. Basic Engng*, Journal of Fluids Engineering (1961).
2. Goldstein, *Classical Mechanics*. Pearson; 3 edition (2001).
3. W. Foundation. Precession. https://en.wikipedia.org/wiki/Rifling .

Chapter 5

2015 Problem 9: Hovercraft

Ruyang Sun[1]*, Heyang Long[2†], Wenli Gao[2], Sihui Wang[2]

[1]*Kuang Yaming Honors School, Nanjing University*
[2]*School of Physics, Nanjing University*

In this paper, the hovercraft model is studied. The relationship between pressure of balloon and its elongation is studied with continuum mechanics model of rubber. The lifting force of hovercraft, which comes from the pressure difference between two sides of the disk is solved by calculating the pressure distribution using Navier-Stokes equation. The parameters affecting floating time of the system are also investigated.

1. Introduction

A simple model can be built using a CD and a balloon filled with air attached via a tube. Exiting air can lift the device making it float over a surface with low friction. Investigate this phenomenon.

Fig. 1.   A simple model hovercraft built with a CD and a balloon.

In 1993, Mihir Sen[4] analyzed the dynamics of a hemispherical dome levitated by a vertical jet of air. He made a lubrication assumption for small separation. In 2010, Charles de Izarra and Gr egoire de Izarra[1]

*E-mail: 141242053@smail.nju.edu.com
†E-mail: 141120065@smail.nju.edu.com

studied the dynamics of a model hovercraft in a more detailed way. They figure out the velocity and pressure distribution. These papers all above didn't verify the stable flow condition and study the floating time. Based on their work, we will verify the steady flow assumption of the balloon and study the parameters which determine the floating time of the hovercraft. The pressure of air in the balloon is larger than the atmosphere pressure outside, so the air jet out of the tube and form a thin air film under the disk. Firstly, we study the relationship between inner balloon pressure and its elongation. Then we measure the flux of the flow and thickness of the air film between disk and floor to determine the of the flow under the disk and verify the constant flow assumption.Then we use the simplified Navier-Stokes equation and some boundary conditions to deduce the pressure and flow velocity of the fluid field beneath the CD.The pressure difference between air under and above the disk provide the lifting force to maintain the floating state. By deriving the expression of lifting force we can find that there exists a stable position for the hovercraft. After we investigating the tube flow, we will study the parameters that influence the floating time of the system.

2. Study on the Inner Air of the Balloon

2.1. *Theory on the Balloon Pressure*

We use the continuum mechanics model of rubber for there exists large deformation ($\lambda_i \geq 2$) in the studied system.[2] Assuming that the rubber is isotropic and homogenous,we have its strain energy density as

$$W = W(I_1, I_2, I_3) \tag{1}$$

where I_1, I_2, I_3 are three invariants of strain tensor, which can be expressed by three principle elongation ($\lambda_1, \lambda_2, \lambda_3$) respectively :

$$\begin{cases} I_1 = \lambda_1^2 + \lambda_2^2 + \lambda_3^2 \\ I_2 = \lambda_1^2\lambda_2^2 + \lambda_2^2\lambda_3^2 + \lambda_1^2\lambda_3^2 \\ I_3 = \lambda_1^2\lambda_2^2\lambda_3^2 \end{cases} \tag{2}$$

Considering the incompressibility of rubber, we have the relation:

$$W = \sum_{i,j=0}^{\infty} C_{ij}(I_1 - 3)^i(I_2 - 3)^j, \tag{3}$$

where i, j is integer. As we control $\lambda_i < 4.5$ in the experiment, this equation can be simplified into:

$$W = C_{10}(I_1 - 3) + C_{01}(I_2 - 3). \tag{4}$$

Assuming that the balloon is a sphere from Fig. 2, we know that there exists a relation,

$$(\pi(R + T)^2 - \pi R^2)\sigma = \pi R^2 p, \tag{5}$$

where σ is the stress tensor, p is the pressure that balloon exerts on the inner air. As $T \ll R$, the pressure p can be expressed as:

$$p = \frac{2T}{R}\sigma \tag{6}$$

whereas the expression of stress tensor is:

$$\sigma = \lambda_1 \frac{\partial W}{\partial \lambda_1} - \lambda_2 \frac{\partial W}{\partial \lambda_2} \tag{7}$$

Since the deformation of the balloon is an equibiaxial tension, the three parameters become:

$$\begin{cases} \lambda_1 = \lambda \\ \lambda_2 = \lambda \\ \lambda_3 = \frac{1}{\lambda^2} \end{cases} \tag{8}$$

And the geometric relation is

$$\begin{cases} T = \frac{t}{\lambda^2} \\ R = \lambda r \end{cases} \tag{9}$$

Combining all the equations above, we can derive the relation between pressure p and elongation λ as:

$$p = \frac{4t}{r}(C_{10} + C_{01}\lambda^2)(\frac{1}{\lambda} - \frac{1}{\lambda^7}) \tag{10}$$

The constants C_{10} and C_{01} are related with the material's property, which can be determined by fitting experimental data and theoretical curve.

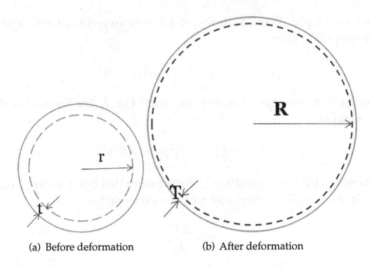

(a) Before deformation   (b) After deformation

Fig. 2.   Deformation of the balloon.

2.2. *Experimental Approach*

Fig. 3 is an experimental set-up designed to determinate C_{10} and C_{01}. First, we fill the balloon with air to the radius R of about four times the original radius r. Then connect it to one side of a one-way valve. The other side of the one-way valve is connected to a U tube barometer which shows the pressure exerted by the balloon. After assembling the apparatus, we let the air leak as slowly as we can from the one-way valve with the balloon shrinking slowly in the meantime. So the process can be considered quasi-static. The pressure and corresponding radius are recorded and compared with the theory(see Fig. 4).

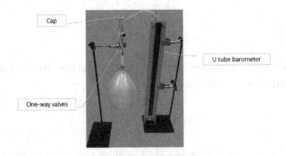

Fig. 3.   Apparatus to measure the pressure of the inner air.

The tension of the rubber when it is stretched and the tension when it shrinks are different according to Mullins Effect. When the hovercraft is floating, the balloon is actually letting air out, so we need to measure the pressure radius relation when the balloon is shrinking rather than expanding.

It is clearly from Fig. 4 that the experimental data is consistent well with the result of the theory (Eq. (10)) and the corresponding constants are $C_{10} = 0.0043$, $C_{01} = 0.98$.

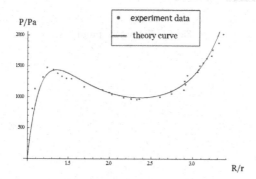

Fig. 4.   Fit the data point with the theoretical curve.

3. Analysis on the Fluid Field Under the Disk

3.1. *The Experiment to Measure Flux of the Flow*

To understand the air in the balloon further, another experiment is designed to find the flux of the flow. In order to measure flux of the flow, a video of the hovercraft exiting air is taken using high-speed camera (Casio EX ZR200) to capture the section of the balloon. Mathematica is used to calculate the volume of the balloon letting out air. Firstly, the function 'EdgeDetect' is used to extract the profile of the section. The pixel points are extracted by the function 'PixelValuePositions'. Half of the pixel points is used to fit a curve $f(x)$. Since the balloon is rotationally symmetric, the volume can be calculated as:

$$V = \int_a^b \pi f^2(x)dx \qquad (11)$$

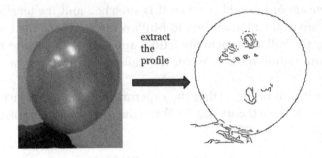

Fig. 5.   Extract the profile.

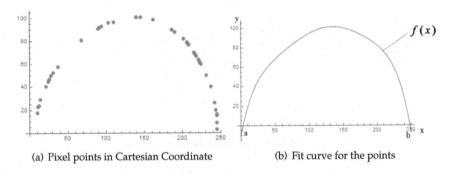

(a) Pixel points in Cartesian Coordinate   (b) Fit curve for the points

Fig. 6.   Calculation of Balloon Volume.

Then a diagram of time and responding balloon volume (see Fig. 6) can be plotted as Fig. 7

In Fig. 7, it is obviously that the air flow of the system is constant, and the flux $Q = 6 \times 10^{-4} m^3/s$ is also calculated.

3.2. *Experiment to Measure the Thickness of the Air Film*

To measure the thickness of the air film, a set-up as shown in Fig. 9 is designed in which a stack of paper is squeezed tightly to let out the air between the layers. Then it is pinned on the table and some glue is attached to the upper surface of it. The thickness of the stack of paper, which can be changed by adding or removing pieces of paper is measured by micrometer. After the hovercraft is released, a little perturbation is given to it, causing it to move on the floor. If the hovercraft is stopped by the stack of paper, the thickness of the stack of paper is larger than

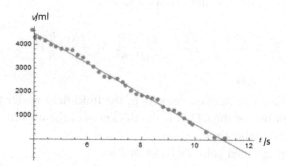

Fig. 7.   The flux of the balloon.

that of air film. If the hovercraft can pass the stack of paper, the result is opposite. The experiment results show that the thickness of air film is 0.5-0.6mm (see Fig. 8).

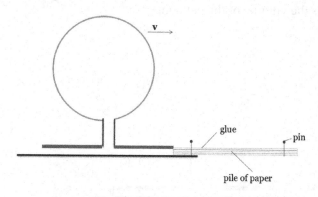

Fig. 8.   The measurement of the thickness of air film.

3.3. *The Determination of Reynolds Number and Simplification*

To study the dynamics of the fluid field, we introduce the Navier-Stokes equation:

$$\rho\frac{\partial \vec{v}}{\partial t} + \rho(\vec{v}\cdot\nabla)\vec{v} = -\nabla P + \mu\triangle\vec{v} \tag{12}$$

To get an approximate analytical solution of this non-linear partial differential equation, the Reynolds Number of the fluid field under the disk has to be investigated and a constant flow assumption has to be made.

The Reynolds Number is defined as below:

$$
\begin{cases}
R^* = \dfrac{\rho(\vec{v} \cdot \nabla)\vec{v}}{\mu \triangle \vec{v}} = \dfrac{\rho U^2/R_1}{\mu U/h^2} = \dfrac{UR_1}{\eta}\left(\dfrac{h}{R}\right)^2 \\
\eta = \mu/\rho
\end{cases}
\tag{13}
$$

where U is the characterized velocity of the fluid field under the disk, R_1 is the outer radius of the CD, h is the thickness of the air film, μ is the dynamic viscosity of the air.

And the characterized velocity is defined as:

$$
U\pi(R_1^2 - R_0^2) = \int_{R_0}^{R_1} v2\pi r dr = \int_{R_0}^{R_1} \dfrac{Q}{2\pi r h} 2\pi r dr
\tag{14}
$$

where R_0 is the inner radius of the tube. Q is the flux, which is defined as:

$$
Q = \dfrac{dV}{dt}
\tag{15}
$$

Where V is the volume of the balloon.

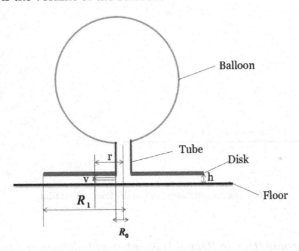

Fig. 9. The sketch of the set-up.

We have measured the flux Q and the thickness of the air film h so the Reynolds Number can be figured out. From Fig. 7 we can know that the constant flow assumption is verified because the flux is constant during the whole process.

Combining the statistics above,The Reynolds number R^* of the flow under the disk is about 0.01, under this condition, the viscous force is far larger than the inertial force, the Navier-Stokes equation can be simplified:[1]

$$\rho\frac{\partial\overrightarrow{v}}{\partial t}+\rho(\overrightarrow{v}\cdot\nabla)\overrightarrow{v} = -\nabla P+\mu\triangle\overrightarrow{v} \Rightarrow 0 = -\nabla P+\mu\triangle\overrightarrow{v} \qquad (16)$$

After the Navier-Stokes equation is simplified, we can derive the fluid distribution of pressure and velocity next.

3.4. *Pressure and Velocity Distribution*

For the pressure the balloon exert on the inner air is much smaller than the atmosphere pressure, the air is taken as incompressible. The continuity equation and Navier-Stokes equation can be written as:

$$\begin{cases} -\nabla P+\mu\triangle\overrightarrow{v} = 0 \\ \nabla\cdot\overrightarrow{v} = 0 \end{cases} \qquad (17)$$

We take the cylindrical coordinate (r,ϕ,z) to study this question, thus the velocity can be written as $\overrightarrow{v} = v_r\overrightarrow{e_r} + v_\varphi\overrightarrow{e_\varphi} + v_z\overrightarrow{k}$. Considering the symmetry of the system, v_φ, v_z=0. So the fluid velocity is along the radial direction.

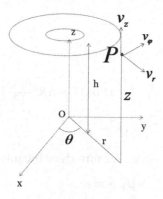

Fig. 10.   Velocity in cylindrical coordinate.

Combining with the boundary conditions:

$$\begin{cases} v_r(z = 0) = 0 \\ v_r(z = h) = 0 \\ p(r = R_1) = P_A \end{cases} \tag{18}$$

And

$$Q = \iint \vec{v_r} \cdot d\vec{S} \tag{19}$$

we can derive the velocity and pressure distribution.

$$P(r) = P_A + \frac{6\mu Q}{\pi h^3} ln(\frac{R_1}{r}) \tag{20}$$

where P_A is the atmosphere pressure.

$$\overrightarrow{v(r,z)} = \frac{3Qz(h-z)}{\pi h^3 r} \vec{e_r} \tag{21}$$

4. Lifting Force and the Stability of the System

The difference between the under pressure and above the CD would provide a lifting force to make the hovercraft float over the floor. The lifting force can be expressed as:

$$F = \int_{R_0}^{R_1} 2\pi r dr + \pi R_0^2 (P(R_0) - P_A)$$

$$\Rightarrow F = \frac{12\mu Q}{h^3} (\int_{R_0}^{R_1} r ln R_1 dr - \int_{R_0}^{R_1} r ln r dr) + \pi R_0^2 \frac{6\mu Q}{h^3} ln(\frac{R_1}{R_0})$$

$$\Rightarrow F = \frac{3\mu Q}{h^3} (R_1^2 - R_0^2) \tag{22}$$

To study the stability of the system, we introduce the total potential,[5]

$$\begin{cases} U_{total} = U_F + mgh, \\ \vec{F} = -\nabla U_F \end{cases} \tag{23}$$

U_F represents the potential of the lifting force F can be obtained from Eq. (4) and the term mgh represents the gravitational potential of the system.

$$U_{total} = \frac{C}{h^2} + mgh, C = \frac{3\mu Q(R_1^2 - R_0^2)}{2} \tag{24}$$

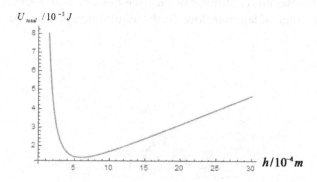

Fig. 11.   The potential curve.

Fig. 11 is the corresponding theoretical curve of U_{total}. As shown in the Fig. 11, we find that the hovercraft will make small oscillation among the stable position and the amplitude of the oscillation will fade away quickly to zero because of the air resistance. The calculation of system air resistance is complicated and the dynamics of the hovercraft is not the point of our research. The existence of a minimum point on the curve implies that there is a stable state of the system, where $h = h_0 = 0.59mm$. Compared with the experimental shown in section 3.2 ($h_0 = 0.5\ 0.6mm$), the theoretical results fit well.

5. The Parameters that Influence the Floating Time

In order to figure out the floating time with fixed balloon volume, we have to study the flux of the flow. It has been verified that the flux of the flow is constant. But from Eq. (4), we know that flux is also related with the stable height(or the thickness of the air film) h, and h can change as the parameters of hovercraft change. The only way to eliminate the stable height h is to establish the fluid-dynamical equation and mechanical equation of the system.

To study the fluid dynamical property of the system, we will study the flow in the tube, the Reynolds number R_* of the tube flow is defined as $R_* = \frac{UD}{\nu}$, U is the fluid velocity at the center of the tube, D is the diameter, ν is the kinetic viscosity. Supposing that the flow is laminar and according to the Poiseuille Law the relationship between mean velocity of the fluid, which is defined as $V = \frac{Q}{S}$ and central fluid velocity is $\frac{V}{U} = \frac{1}{2}$ so we calculate the Reynolds number of the tube flow R_* is 2000-3000, which is within the range of laminar flow. So the assumption is right. Based on the

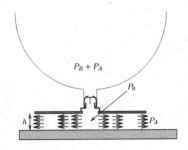

$$P_B + P_A$$
$$P_0$$
$$h$$
$$P_A$$

Fig. 12.   The laminar flow of the system.

Poiseuille Law

$$Q = \frac{\pi D^4 \triangle p}{128 \mu L} \tag{25}$$

and Eq. (20), we have

$$Q = \frac{\pi D^4 h^3 L P_B}{\mu(128 h^3 L + 6 D^4 ln(R_1/R_0))} \tag{26}$$

The Eq. (26) is the fluid dynamical equation. where P_B is the pressure exerted by the balloon. From our previous research on the pressure exerted by the balloon in section 1, we find P_B is maintained within a small range($1000 \pm 100 Pa$) during most of floating time. So we can simplify the question by taking P_B as a constant value.

Combining the Eq. (26) and Eq. (4) which is the mechanical equation of the system, we can eliminate the stable height h and derive the relationship between floating time and the mass of the system

$$mg = K_1 - K_2 \frac{1}{t}, K_1 = \frac{\pi P_B(R_1^2 - R_0^2)}{2ln(R_1/R_0)} K_2 = \frac{4\mu V_0(R_1^2 - R_0^2)L}{R_0^4 ln(R_1/R_0)}$$

where V_0 is the initial volume of the balloon, L is the length of the tube. To investigate the relationship between mass of hovercraft and floating time, we set the initial volume of the balloon and other geometric parameters fixed and load the hovercraft symmetrically with mass (Fig. 13) to prevent the inclination of the disk.Then we measure the floating time of hovercrafts with different mass and the obtained relation is plotted in Fig. 14.

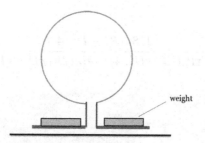

Fig. 13. Put weight symmetrically on the disk.

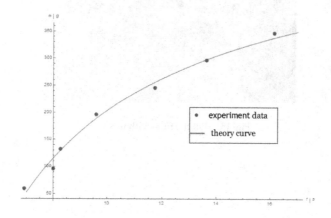

Fig. 14. The mass-time relation.

Fig. 14 reveals that if we load more weight on the hovercraft, it can float over the floor for a longer time. However from Eq. (24), it is clear that more weight on the hovercraft can cause the stable position to lower, which increases the risk that the hovercraft will hit the floor if it is not released in a proper height.

The floating time is also concerned with the inner diameter of the tube. In order to investigate this relationship, a set-up is designed as Fig. The glass glue and paraffins are mixed to fill the tube to ensure a smooth inner wall while a rough inner wall which consists different boundary conditions may cause the flow to become unstable. With other geometric size and the mass of the system unchanged, we can change the inner diameter of the tube. Then we measure the floating time of hovercrafts with different inner diameters. We can derive the relation by rewrite the Eq. (27).

$$t = V_0 \frac{128\mu(R_1^2 - D^2/4)L}{\pi P_B(R_1^2 - D^2/4) - 4ln(2R_1/D)mgD^4} \tag{27}$$

Fig. 15.   Fill the tube with mixture.

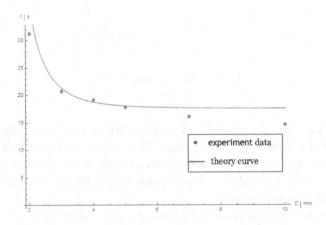

Fig. 16.   The relationship between diameter of the tube and time.

In the Fig. 15 and Fig. 16, we can notice that when the diameter of the tube is less than 0.5cm, the experimental data points fit with the theoretical curve, whereas when the diameter becomes larger the experimental results deviate a lot from the theory. that's because the flow will be gradually transferred from laminar flow into turbulent flow and the system will oscillate with larger and unpredictable amplitude,[3] so in the experiment, we should try to avoid this condition.

6. Conclusion

By measuring the inner pressure of a balloon and calculating the Reynolds Number of the system, we find that the flow within the system is laminar. With the simplified Navier-Stokes Equation, we predict theoretically the relation of floating time to various parameters such as inner radius of the tube and the total mass of the system. The prediction fit well with the experiments conducted later. However, there are still many undetermined areas for research regarding this topic. For example, the properties of the surface on which the hovercraft is 'sliding' and the velocity of the air flow can affect the floating time but haven't been studied in my research for the limit of apparatus and theory.

References

1. Charles De Izarra and Gregoire De Izarra. Stokes equation in a toy cd hovercraft. *European Journal of Physics*, 32(1):89–99, 2011.
2. D. R. Merritt and F. Weinhaus. The pressure curve for a rubber balloon. *American Journal of Physics*, 46(10):976–977, 1978.
3. Herrmann Schlichting and Klaus Gersten. *Boundary-Layer Theory*. McGraw-Hill,, 1960.
4. Mihir Sen. Dynamic analysis of a hemispherical dome levitated by an air jet. *Applied Mathematical Modelling*, 17(5):226–235, 1993.
5. Charles Thompson. Stability of a stokes boundary layer. *Journal of the Acoustical Society of America*, 81(4):861–873, 1987.

Chapter 6

2015 Problem 10: Singing Blades of Grass

Dachuan Lu[1*], Yiran Deng[2†], Boyuan Tao[1], Bingnan Liu[1], Wenli Gao[3]

[1]*Kuang Yaming Honors School, Nanjing University*
[2]*School of Earth Science and Engineering, Nanjing University*
[3]*School of Physics, Nanjing University*

In this study, acoustic properties of the pure tone produced by a vibrating plate with in-plane tension in a uniform parallel flow is investigated theoretically and experimentally. By modifying Euler-Bernoulli beam equation combined with the mean flow theory, an explicit expression for the frequency of the pure tone is derived using eigenvalue method, in which all of the parameters can be measured in experiments. To verify the new equation, experiments have been designed to investigate the effects of the flow velocity and in-plane tension on the frequency and intensity of the sound, whose results have a good agreement with the theoretical results.

1. Introduction

It is possible to produce a sound by blowing across a blade of grass, a paper strip or similar. Investigate this effect.

It is possible to produce a sound when we blow across a blade of grass in daily life. This phenomenon is caused by interactions between structures and the axial uniform flow, which is also encountered in industrial fields. For example, the cross-sea bridge may vibrate due to the strong wind and cause large damage. This intriguing and practical issue was first gained attention from Rayleigh in 1879, who considered a tension-free surface of infinite length vibrating in the flow.[1] In 1988, Dowling investigated the dynamics of towed under-water cables, which is concerning about the effect of tension.[2] Paidoussis systematically studied the general

---

*E-mail: dclu@smail.nju.edu.cn
†E-mail: dengyiran_nju@163.com

slender structures in axial flow and the properties of their vibration in a mechanical way.[3] Morris-Thomas studied the stability and the drag of a flexible sheet under in-plane tension in uniform flow, and gave precise results of modeling the flag.[4] Clark turned attention to acoustics aspect.[5] He investigated the aeroelastic structural acoustic coupling, concerning about the noise transmission through the panel.

However, the acoustical mechanism and characteristics of the sound produced by the plates vibrating in uniform flow with in-plane tension have hitherto not been studied, and it is an interesting problem in daily life but with deep underlying physics.

In this paper, we develop a practical model to investigate the dependence of the frequency, the intensity and the spectrum of the sound produced by the plate under the different conditions.

2. Model Development

In order to understand the effect of the uniform flow on the vibrating plates, several assumptions should be given.

(1) The plate is immersed in a uniform flow coming from positive x-axis.
(2) For a large Reynold number, the viscosity of the fluid can be neglected.
(3) The cross flow of a plate is at a small angle of attack and the ambient air is approximately incompressible.

Here we mainly consider rectangle plates. Fig. 1 shows the geometry and corresponding parameters of the model. The flow comes out from a tube with a velocity of **u** parallel to the surface of the rectangle plate. H_1, H_2 are the thickness of the cross flow, where the extremely strong interaction between the plate and the ambient air makes it incompressible. However, the air in the outer region is compressible to propagate the sound. b and L are the width and length of the plate respectively.

In a qualitative view, as there exist turbulent components in the flow, it will lead the plate to vibrating, exciting the motion of the ambient air. In turn, the variation of ambient pressure will affect the vibration of the plate. Hence, the interaction between the plate and the ambient air produces a pure tone.

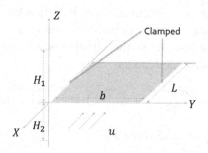

Fig. 1. The geometry sketch and the notations of the model.

Based on the above assumptions and analysis, we will give a theoretical model to describe the interactions between the plate and the ambient air.

2.1. *Free Vibration of the Plate with In-plane Tension*

For simplification, we consider the plate is homogeneous and isotropic. Based on the 2-dimension generalized Euler-Bernoulli beam equation,[6] and assuming that the tension per unit length T distributes homogeneously along the y-direction in the plate, the dynamic equation for the free vibration of the plate can be written as,

$$D\left(\frac{\partial^4}{\partial x^4} + 2\frac{\partial^2}{\partial x^2 \partial y^2} + \frac{\partial^4}{\partial y^4}\right)w - T\frac{\partial^2 w}{\partial y^2} + \mu\frac{\partial^2 w}{\partial t^2} = 0 \qquad (1)$$

where $w(x,y)$ describes the deflection of the plate in z direction, at position (x,y), μ is the mass of the plate per unit area, $D = \frac{Eh^3}{12(1-\sigma^2)}$ defines the flexural rigidity, E is Young modulus, h is the thickness of the plate and σ is Poisson ratio.

The boundary conditions associated to Eq. (1) can be written as:

$$\begin{cases} w\Big|_{y=\pm\frac{b}{2}} = 0, \frac{\partial w}{\partial y}\Big|_{y=\pm\frac{b}{2}} = 0 \\ \frac{\partial^2 w}{\partial x^2}\Big|_{x=\pm\frac{L}{2}} = 0, \frac{\partial^3 w}{\partial x^3}\Big|_{x=\pm\frac{L}{2}} = 0 \end{cases} \qquad (2)$$

respectively for the clamped side parallel to the x-axis and free end parallel to the y-axis, as Fig. 1 illustrates.

To solve Eq. (1), combined with Eq. (2), we can make an ansatz, $w(x, y, t) = w_0 \exp\left(k_x x - i\frac{m\pi}{b}y - i\omega t\right)$, where w_0 is the vibration amplitude, k_x

is the wave number in x direction. Since in y direction, the two sides are clamped, m describes the stationary wave number in y direction, between the two clamped sides. ω is the circular wave frequency. Then the dispersion equation can be read as:

$$D\left(k_x^4 - \frac{2k_x^2 m^2 \pi^2}{b^2} + \frac{m^4 \pi^4}{b^4}\right) + \frac{m^2 \pi^2 T}{b^2} - \mu\omega^2 = 0 \tag{3}$$

Rewrite the coefficients and solve it for k_x,

$$k_x^4 - 2\beta^2 k_x^2 - \gamma^4 = 0 \tag{4}$$

where $\beta = \frac{m\pi}{b}$ and $\gamma^4 = \frac{\mu\omega^2 - \beta^2 T}{D} - \beta^4$. Eq. (4) has four roots for k_x, namely:

$$\begin{cases} k_x = \pm k_{x1}, k_{x1}^2 = \sqrt{\beta^4 + \gamma^4} + \beta^2 \\ k_x = \pm i k_{x2}, k_{x2}^2 = \sqrt{\beta^4 + \gamma^4} - \beta^2 \end{cases} \tag{5}$$

According to Vieta's formulas, there exist,

$$\begin{cases} k_{x2}^2 = k_{x1}^2 - 2\beta^2 \\ k_{x1} k_{x2} = \gamma^2 \end{cases} \tag{6}$$

So, we write the general solution as:

$$\begin{aligned} w = &\exp\left(-i\tfrac{m\pi}{b}y - i\omega t\right) \\ &\times \left(A\cosh(k_{x1}x) + B\sinh(k_{x1}x) + C\cos(k_{x2}x) + D\sin(k_{x2}x)\right), \end{aligned} \tag{7}$$

Substituting the general solution into the boundary conditions (Eq (2)), we can determine k_x. First, we consider the even functions in general solution:

$$\begin{cases} k_{x2}^2 C \cos\left(k_{x2}\tfrac{L}{2}\right) = k_{x1}^2 A \cosh\left(k_{x1}\tfrac{L}{2}\right) \\ k_{x2}^3 C \sin\left(k_{x2}\tfrac{L}{2}\right) = -k_{x1}^3 A \sinh\left(k_{x1}\tfrac{L}{2}\right) \end{cases} \tag{8}$$

combined with Eq. (6), Eq. (8) can be written as:

$$\tan\left(k_{x2}\frac{L}{2}\right) = -\sqrt{1 + \frac{2\beta^2}{k_{x2}^2}} \tanh\left(\sqrt{k_{x2}^2 + 2\beta^2}\,\frac{L}{2}\right) \tag{9}$$

k_{x2} can be solved numerically according to Eq. (9). If k_{x2} is large enough, the equation has a asymptotic form, $k_{x2} = \frac{2}{L}\left(-\frac{1}{4} + n\right)\pi$, where n is a large enough integer.

Solving the dispersion equation Eq. (3) for the circular wave frequency ω, we arrive at:

$$\omega = \sqrt{\frac{D\left(k_{x2}^2 + \beta^2\right)^2 + \beta^2 T}{\mu}} \tag{10}$$

As k_{x2} has a series of solutions due to the periodicity of tangent, ω should have a corresponding series of values. In the even function case, the series of solution ω can be labelled as, $\omega_2, \omega_4, \omega_6...$

Similarly, we can deduce the odd sequence of ω by solving the following equation for k_{x1}:

$$\tan\left(k_{x1}\frac{L}{2}\right) = \sqrt{\frac{k_{x1}^2}{k_{x1}^2 + 2\beta^2}} \tanh\left(\sqrt{k_{x1}^2 + 2\beta^2}\frac{L}{2}\right) \tag{11}$$

Using Eq. (10) to find $\omega_1, \omega_3, \omega_5...$

So far, we have figured out the eigen frequency of free vibration. Above all, the spatial wavenumbers have been derived and can be used in the situation described in the following section, because the spatial wavenumbers remain unchanged among different loads.

2.2. *The Effect of Uniform Flow*

Considering the influenced pressure variation Δp as a external load, the correponding dynamic equation should be:

$$D\left(\frac{\partial^4}{\partial x^4} + 2\frac{\partial^2}{\partial x^2 \partial y^2} + \frac{\partial^4}{\partial y^4}\right)w - T\frac{\partial^2 w}{\partial y^2} + \mu\frac{\partial^2 w}{\partial t^2} = \Delta p \tag{12}$$

For the inviscid and incompressible fluid, we can use potential flow theory to calculate the fluid motion. Since the upper and the lower region of the plate is disconnected, we should analyze the fluid motion respectively. Write the velocity potential in the following form:[3]

$$\Phi_{0j} = U_j x + \Phi_j(x,y,z,t), j = 1,2 \tag{13}$$

The first item in RHS represents the uniform flow with a velocity of U_j, while the last item represents the perturbation potential. $j = 1,2$ represents the upper and lower region of the plate. Hence, the velocity field for the perturbed state is given by

$$v_{jx} = U_j + \frac{\partial\Phi_j}{\partial x}, v_{jy} = \frac{\partial\Phi_j}{\partial y}, v_{jz} = \frac{\partial\Phi_j}{\partial z} \tag{14}$$

Similarly, the pressure on the surface of the plate can be written as

$$p_{0j} = p_j^0 + p_j \tag{15}$$

Where p_j^0 corresponds to the steady potential flow, and p_j represents the perturbed component.

The governing equations are the incompressible and inviscid Navier-Stokes equation together with the continuity equation and the boundary conditions. The perturbation potential firstly satisfies the continuity equation and can be expressed as:

$$\nabla^2 \Phi_j = 0 \tag{16}$$

The boundary conditions can be classified into two parts. One is the impermeability condition, which means the fluid cannot get into the plate. Thus, the velocity of air elements should be consistent with the plate. It can be expressed as

$$\frac{\partial \Phi_j}{\partial z}\Big|_{z=z_w} = \frac{\partial w}{\partial t} + U_j \frac{\partial w}{\partial x} \tag{17}$$

where $z_w = \pm \frac{h}{2} + w(x, y, t)$ represents the position of the plate surface. Since both h and w are extremely small($h, w \ll H_j$), an approximation can be made, $z_w = \pm \frac{h}{2} + w(x, y, t) \simeq 0$. This boundary condition is of significance, which is the key to link the interaction between the plate and ambient air.

The other part of boundary conditions is

$$\begin{cases} \frac{\partial \Phi_1}{\partial z}\Big|_{z=-H_1} = 0, \\ \frac{\partial \Phi_2}{\partial z}\Big|_{z=H_2} = 0 \end{cases} \tag{18}$$

As we set the stiff boundary at $z = -H_1$ and $z = H_2$, the normal velocity there is always zero.

Because of the interaction between the plate and the air, the motion of ambient air is similar to the vibration of the plate and only differs in a phase. We can write the velocity potential perturbation in form of

$$\Phi_j = Z_j(z) \exp\left(k_x x - i\frac{m\pi}{b}y - i\omega t\right). \tag{19}$$

Substituting Eq. (19) into the impermeability condition, Eq. (17), we have:

$$Z_j'(0) \exp\left(k_x x - i\frac{m\pi}{b}y - i\omega t\right) = \frac{\partial w}{\partial t} + U_j \frac{\partial w}{\partial x} \tag{20}$$

So, Φ_j is read as:

$$\Phi_j = \frac{Z_j(z)}{Z_j'(0)}\left(\frac{\partial}{\partial t} + U_j\frac{\partial}{\partial x}\right)w \qquad (21)$$

Where $Z_j(z)$ can be found by solving the Eq. (16) and Eq. (18).
 Considering the Bernoullis equation:

$$\frac{\partial \Phi}{\partial t} + \frac{V^2}{2} + \frac{p}{\rho} = 0 \qquad (22)$$

where $V^2 = v_x^2 + v_y^2 + v_z^2$, the pressure on the surface p_{0j} is given by:

$$p_{0j} = -\rho\left(\frac{\partial \Phi_j}{\partial t} + \frac{1}{2}\left(U_j + \frac{\partial \Phi_j}{\partial x}\right)^2 + \left(\frac{\partial \Phi_j}{\partial y}\right)^2 + \left(\frac{\partial \Phi_j}{\partial z}\right)^2\right) \qquad (23)$$

Substituting Eq. (15) into Eq. (23), the perturbed component is read as:

$$p_j = -\rho\left(\frac{\partial \Phi_j}{\partial t} + U_j\frac{\partial \Phi_j}{\partial x}\right) \qquad (24)$$

Substituting Eq. (21) into Eq. (24), we arrive at

$$p_j = -\rho\frac{Z_j(0)}{Z_j'(0)}\left(\frac{\partial}{\partial t} + U_j\frac{\partial}{\partial x}\right)^2 w \qquad (25)$$

Assuming $U_j = U$, the variation of pressure is,

$$\Delta p = p_2 - p_1 = -\rho(k_x U - i\omega)^2 \left(K_{p1} + K_{p2}\right) \qquad (26)$$

where $K_{pj} = \dfrac{\coth\left[\sqrt{-k_x^2 + \frac{m^2\pi^2}{b^2}}H_j\right]}{\sqrt{-k_x^2 + \frac{m^2\pi^2}{b^2}}}, j = 1, 2.$

k_x, m remain unchanged, for the plate's boundary conditions here are the same as those of the free vibration.
 Considering $k_x = \pm ik_{x2}$, the total dispersion equation becomes:

$$D\left(k_x^2 + \beta^2\right)^2 + \beta^2 T - \mu\omega^2 = -\rho(k_x U + \omega)^2 \left(K_{p1} + K_{p2}\right) \qquad (27)$$

By solving Eq. (27), we can get the circular wave frequency ω as:

$$\omega^2 = a_U U^2 + b_T T + c \qquad (28)$$

where $a_U = \dfrac{k_x^2\mu\rho\left(K_{p1}+K_{p2}\right)}{\left(\mu+\rho\left(K_{p1}+K_{p2}\right)\right)^2}, b_T = \dfrac{\beta^2}{\mu+\rho\left(K_{p1}+K_{p2}\right)}, c = \dfrac{D\left(k_x^2+\beta^2\right)^2}{\mu+\rho\left(K_{p1}+K_{p2}\right)}.$

2.3. *Sound Intensity*

The sound intensity Π can also be calculated using the following formula[6]

$$\Pi = \frac{1}{2} \Re A \frac{|p|^2}{c\rho} \tag{29}$$

where p is pressure, A is area of the plate, c is the ambient sound velocity, ρ is the density of ambient air.

Substituting Eq. (26) into Eq. (29), sound intensity can be expressed as:

$$\Pi = \frac{Aw_0^2\rho(k_xU + \omega)^4 \coth\left(H\sqrt{k_x^2 + \beta^2}\right)^2}{2c\left(k_x^2 + \beta^2\right)} \tag{30}$$

keeping other parameters constant, it can be found that the sound intensity Π is proportional to U^4.

Based on the theoretical analysis, a comparative experiment has been designed as follows.

3. Experiment Arrangement

A simple and efficient apparatus displays in the Fig. 3 and Fig. 2. The filmy rectangle plate is made of PVC, with a thickness of $h = 5.7 \times 10^{-5}m$, length $b = 2.24 \times 10^{-2}m$, width $L = 1.18 \times 10^{-2}m$, mass per unit $\mu = 0.09\text{kg/m}^2$, and the Youngs Modulus measured by definition is $E = 4.536 \times 10^8\text{Pa}$, Poisson ratio $\sigma = 0.31$ supplied by the manufacturer.

Fig. 2.   The set-up of the apparatus.

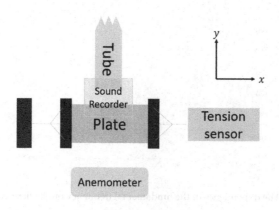

Fig. 3.   Sketch of the apparatus.

Stiff rubber with flat surface is used to clamp the plate which ensures the tension in plane and along y direction. The tension is measured by PASCO Tension Sensor in equilibrium position. The nozzle of the tube is larger than the size of the plate, so that the edge effects can be ignored. Flow velocity is measured by anemometer (GM8901) before processing the experiment. The plate is set at where we measure the flow velocity. Considering the effects of plate's usage history, namely Mullins effect,[7] the plate used in experiment has been stretched many times.

Sound was recorded by recorder (Sony), and the frequency spectrum was analyzed by professional software (Cool edit). Sound intensity is measured by the PASCO sound sensor.

4. Results

As theory analysis illustrated in section 2.3, to clarify the relation, we rewrite Eq. (28) as $\omega = \sqrt{a_U U^2 + c_U}$, where $c_U = b_T T + c$ are independent of U. We find that if U is large enough, $\omega \propto U$, and that if $U \to 0$, the frequency degenerates to the frequency of free vibration, say, the sound produced by plucking the plate.

Fig. 4 shows the relationship between the frequency and the flow velocity, The theoretical curve is obtained by substituting all the parameters of the plate into Eq. (28), compared with the theoretical curve, the experimental data have a good agreement at large flow velocity.

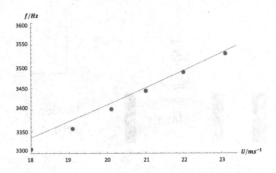

Fig. 4.   The dependence of the fundamental frequency on the flow velocity.

Similarly, Eq. (28) can be written as $\omega = \sqrt{b_T T + d_T}$, where $d_T = a_U U + c$ are independent of tension T. And it hints that the vibration of the plate is akin to the relation in string vibration, where $\omega \propto \sqrt{T}$. Fig. 5 illus-

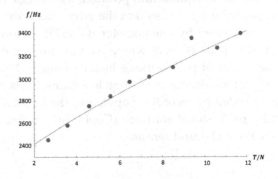

Fig. 5.   The dependence of the fundamental frequency on the in-plane tension.

trates the influence of the in-plane tension on the fundamental frequency. The experiment data have a good agreement with the theoretical values.

Some other relationships have been investigated qualitatively. The corresponding results have been listed in Table 1.

Table 1.   Some other relationships.

Parameters	Correlation with fundamental frequency
Length (L)	Negative correlation
Width (b)	Negative correlation
Mass per unit area (μ)	Negative correlation
Thick of air layer (H)	Positive correlation

And the Fig. 6 illustrates the relationship between sound intensity and flow velocity. The flow velocity has been increased in a uniform velocity, and sound sensor is used to measure the sound intensity. Because the resolution of the sensor is very high, it demonstrates the waveform. In the Fig. 6, the sound amplitude signal has been squared, and only the envelope of the data counts. The both tendencies are consistent. It is notable that when $U > 30m/s$, it outstrips the valid range of the pump.

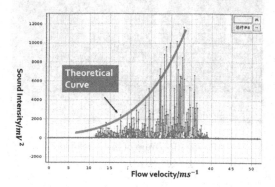

Fig. 6. Sound intensity have positive correlation to flow velocity.

Harmonic is defined as a wave with a frequency that is an integer multiple of the fundamental frequency, and can be found in the spectrum of the plate produced sound (Fig. 7). And the expression for the frequency is $f_n = nf_0$, where n is an integer. However, as Fig. 8 illustrates, there exists another series of frequency possessing unique property, which can

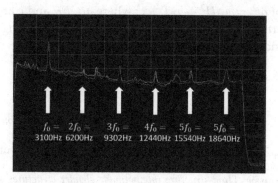

Fig. 7. Spectrum of the pure tone (purple and blue represent left and right channel respectively).

be expressed as: $f_n = (n + a)f_0$, where f_0 is the fundamental frequency, and n is an integer, meanwhile a is a constant between 0 and 1. This phenomenon accounts for the shifts of the wave number in x direction.

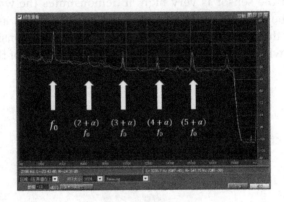

Fig. 8.   Spectrum of the pure tone.

5. Conclusion

In this study, we investigate the acoustic properties of a vibrating plate in the uniform flow by analyzing the interaction between the plate and the ambient potential flow. By combining the Euler-Bernoulli beam theory and mean flow theory, we derive a precise expression (Eq. (27)) for calculating the frequency of the pure tone produced by a plate in uniform flow. Experimental results have a good agreement with the theoretical predictions of $\omega^2 = a_U U^2 + b_T T + c$, where a_U, b_T, and c are dependent on the properties of the plate and the ambient air, indicating the validity of the theoretical model.

References

1. J. W. Strutt and L. Rayleigh, On the instability of jets, *Proc. London Math. Soc.* **10**, 4–13 (1878).
2. A. Dowling, The dynamics of towed flexible cylinders part 1. neutrally buoyant elements, *Journal of Fluid Mechanics.* **187**, 507–532 (1988).
3. M. P. Paidoussis, *Fluid-structure interactions: slender structures and axial flow.* vol. 1, Academic press (1998).
4. M. T. Morris-Thomas and S. Steen. The response of a flexible sheet under axial

tension immersed in parallel flow. In *ASME 2008 27th International Conference on Offshore Mechanics and Arctic Engineering*, pp. 737–744 (2008).

5. R. L. Clark and K. D. Frampton, Aeroelastic structural acoustic coupling: Implications on the control of turbulent boundary-layer noise transmission, *The Journal of the Acoustical Society of America*. **102**(3), 1639–1647 (1997).

6. L. E. Kinsler, A. R. Frey, A. B. Coppens, and J. V. Sanders, Fundamentals of acoustics, *Fundamentals of Acoustics, 4th Edition, by Lawrence E. Kinsler, Austin R. Frey, Alan B. Coppens, James V. Sanders, pp. 560. ISBN 0-471-84789-5. Wiley-VCH, December 1999*. **1** (1999).

7. J. Diani, B. Fayolle, and P. Gilormini, A review on the mullins effect, *European Polymer Journal*. **45**(3), 601–612 (2009).

Chapter 7

2015 Problem 11: Cat's Whisker

Zengquan Yan*, Hanqi Feng, Yuehui Li, Yu Zhang, Chunfeng Hou,
Yuxiao Wang

School of Physics, Harbin Institute of Technology

1. Introduction

The first semiconductor diodes, widely used in crystal radios, consisted
of a thin wire that lightly touched a crystal of a semiconducting material
(e.g. galena), see Fig. 1. Build your own 'cat's-whisker' diode and
investigate its electrical properties.

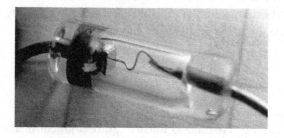

Fig. 1.   Cat's Whisker

The characteristics of the semiconductors were first found by Michael
Faraday in 1833.[1] In 1874, German scientist Karl Ferdinand Braun found
metal sulfides' unilateral conductivity and built a radiodetector.[2] The
invention of radiodetector promotes the development of early wireless
communication. With the PN junction theory established by W.B.Shockley
in 1950,[3] various diodes with different functions were applied in produc-
tion, which lays a ground work for early electronic devices.

*E-mail: 2726588317@qq.com

In this article, a cat's whisker diode is built to reproduce the structure of a semiconducting crystal diode. A small area near the contact point between metal wire and semiconducting crystal forms a PN junction, thus the cat's whisker becomes a diode. The main factors to influence the electrical properties are identified as the position of the contact point, the pressure upon the contact point, the materials of the wire and semiconducting crystals. By using PbS, known as galena, as the base to create cat's whisker, this element has the same electrical properties as diodes. When the base is replaced with SiC, our cat's whisker not only has the usual electrical properties as a diode, but also gives out weak blue light like an LED. In order to investigate cat's whisker's electrical properties, we increase the positive voltage beyond the cat's whisker's working range to perform extreme experiment.

2. Principle of the Diode

The contact between different types of semiconductors forms a PN junction. There is a space-charge region between p-doped region and n-doped region, which makes PN junction unilateral conductive. A PN junction also has junction capacity, composed of diffusion capacity and barrier capacity. The expression of diffusion capacity is[4]

$$C_d = \frac{Aq^2}{KT} \exp\left[\frac{q(V - V_D)}{KT}\right] (L_n N_D + L_p N_A)$$

PN junction can be classified as linearly graded junction, abrupt junction and unilateral abrupt junction. In this problem, a cat's whisker is made of a semiconducting crystal and a metal wire. It contains an alloy junction, which is a shallowly diffused junction with high surface concentration. The junction of a cat's whisker can be treated as unilateral abrupt junction. The expression of barrier capacity is[4]

$$C_T = A\sqrt{\frac{\varsigma_r \varsigma_0 q N_B}{2(V_D - V)}}$$

The total junction capacity is $C = C_d + C_T$.

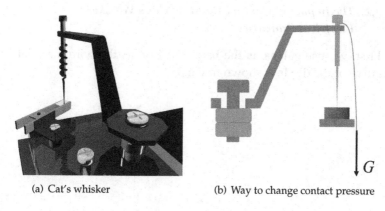

| (a) Cat's whisker | (b) Way to change contact pressure |

Fig. 2. The structure of a Cat's Whisker.

3. Experiment Design

3.1. *Design of a Cat's Whisker*

In making this device, we use alligator clip to hold the base crystal. An aluminum supporting bracket shown in Fig.2(a) with a screw on the top, which can adjust the height. A spring shaped wire is fastened on the screw to ensure that the thin wire can reach the surface of the crystal. A protective resistance is added into the circuit in order to prevent the crystal from burning out. All of the above apparatus are placed on a piece of thick plexiglass board and fixed on a heavy steel plate to ensure that the contact point won't be moved by other factors.[5]

The crystal holder and the aluminum support bracket is revolvable around their own rotation axis. Take the tip of the thin wire as a point and the crystal as an area. The contact point can be changed by rotating the crystal holder and the support bracket.

In order to control the effect of contact pressure, the force upon the contact point should be changeable. In Fig.2(b), a thin rope is hang on the support bracket, by adding different weights on the rope to change the pressure of the contact point. The galena crystal used in the experiment is nature galena, lead gray, metallic luster, Mons' hardness 2.5, density 7.4-7.6g/cm^3. Silicon carbide used in the experiment is also nature crystal, Mons' hardness 9.5, density 3.20-3.25g/cm^3.

3.2. The Impact of Contact Point on Cat's Whisker's Electrical Properties

First, we use galena as the base and iron as the wire material. The I-V curve of the diode is shown in Fig.3.

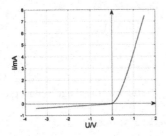

Fig. 3.　PbS base, Fe wire diode I-V curve.　　　Fig. 4.　1N4148 Commercial diode I-V curve.

Comparing Fig.3 and Fig.4, the Cat's Whisker we built has the similar shape with the commercial diode (1N4148), but commercial diode's forward resistance is much lower than cat's whisker's forward resistance and commercial diode's reverse resistance is much larger than cat's whisker's reverse resistance. So the cat's whisker we built does have the electrical properties of a diode, but its' performance is inferior to the commercial diode.

For PbS base crystal and Fe wire, and set the contact pressure as constant, we change the contact points for comparison. I-V curves with three different qualities are shown in Fig. 5 and Fig. 6.

Different contact points change the device's qualities as well as the I-V curves. The difference between them are too obvious to ignore. All experiment data couldn't be measured in one point. But if different contact points have close qualities, the difference between their I-V curves can be ignored.

After repeated measurements, we found that most of the I-V curves for different contact points are linear, which means they are pure resistance-type. Most of the contact points are insufficient in quality. A small part of them are good enough to conduct further experiment. Only "good" contact points can produce a good I-V characteristic. The main reason for this phenomenon is that the crystal we use in experiments is nature crystal and the components in nature crystal are inhomogeneous. The following experiments will be conducted under good contact condition.

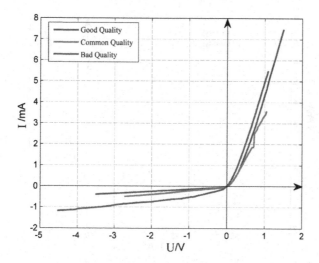

Fig. 5. I-V curves for different contact points with different qualities (PbS base, Fe wire).

In order to conduct further experiments, we need to identify the quality of the diode based on its characteristics of half-wave rectification. Connect the circuit as shown in Fig. 7, applying sine voltage and monitoring the resistance's voltage on an oscilloscope. The theoretic signal simulated by Multisim is also shown in Fig. 7.

The closer the measured rectification signal is to the theoretic result, the better quality the diode is. In this way, the quality of the contact point can be identified immediately.

3.3. *Impact of Contact Pressure*

The pressure of contact point has great impact on the electrical properties of the diode. By hanging weights on the screw, we change the pressure upon the contact point. Other factors such as the materials of the crystal and the wire, input signal frequency and voltage, position of contact point are all fixed.

According to Fig. 8, as the increase of the pressure upon the contact point, ratio of output forward peak voltage value and half of the input peak-to-peak voltage value increases steadily, which also means that the quality of the cat's whisker decreases as the pressure increases on the contact point. The reason is that we can't ignore the junction capacitance

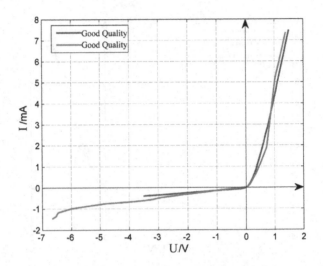

Fig. 6.   I-V curves for different contact points with good qualities (PbS base, Fe wire).

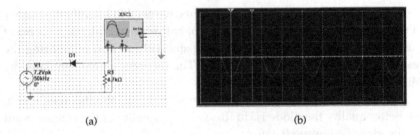

(a)   (b)

Fig. 7.   Half-wave rectification circuit and the result simulated by Multisim.

of the cat's whisker. As can be seen in the expression.

$$C = C_d + C_T = \frac{Aq^2}{KT} \exp\left[\frac{q(V - V_D)}{KT}\right](L_nN_D + L_pN_A) + A\sqrt{\frac{\xi_r\xi_0 qN_B}{2(V_D - V)}}$$

Both diffusion capacity and barrier capacity are proportional to contact area A. The increasing of the contact pressure causes the increasing of contact area, so that the junction capacity C is increased. Because the input signal is a sine function signal with constant frequency, the increase of junction capacity left little time for diode to turn off, which makes the diode lose the ability of unilateral conduction.

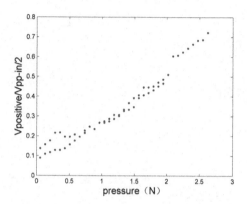

Fig. 8.   PbS base, Fe wire diode, the same contact point with different pressure.

In conclusion, under constant conditions of contact position, materials of the semiconducting crystal and the thin wire, the quality of diode decreases with the increase of pressure upon contact point.

3.4. *Impact of the Wire's Material*

Material of the thin wire may have a great impact on the diode's electrical properties.

The formation mechanism of alloy junction is the injection of metal ions into the semiconducting material and creates a shallow diffused junction. Most of the space charge region is in the semiconducting material. So the main electrical properties of diode is determined by the nature of its' base, not the material of the thin wire. The electrical properties of a diode with the same base and different material of thin wire should be almost the same to each other.

We choose three most common metals to make the thin wire: copper, iron and aluminum. The results are shown in Fig. 9 and Table 1.

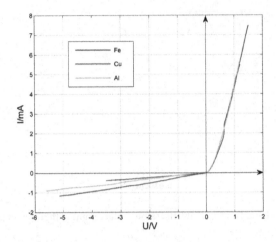

Fig. 9.   I-V curves for diodes with PbS base and wire of three different metals.

Table 1.   Diodes with PbS base and three different metals.

Metal Material	Cu	Fe	Al
Reverse breakdown voltage/V	12.4	12.4	12.4
Inverse current/mA	$\leqslant 2$	$\leqslant 1$	$\leqslant 2$
Max average forward rectified current/mA	5	5	5

From these results, we come to the conclusion that under the same conditions of contact quality, contact pressure and material of the semi-conducting crystal, the electrical properties of the diode doesn't change with the material of its' thin wire.

3.5. *Impact of the Material of the Semiconductors*

Material of the semiconductors may also have a great impact on the diode's electrical properties. We choose two different crystal galena(PbS) and silicon carbide(SiC).

The crystal surface of silicon carbide is irregular, because nature silicon carbide crystal is a combination of different crystal forms. So it is harder to find a good quality position on the surface of a silicon carbide.

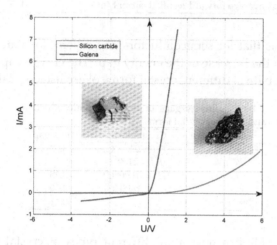

Fig. 10. I-V curves for diodes with galena base and silicon carbide base, Fe wire.

I-V curves of silicon carbide and galena base are similar in shape. If we put these two curves in one figure, we see when $U \geqslant 0$, the forward resistance of the silicon carbide base diode is bigger than that of a galena base diode, which means the quality of a silicon carbide base diode is not as good as that of a galena base diode. When $U < 0$, galena base diode's resistance is smaller than the resistance of a silicon carbide base diode, which means the quality of a silicon carbide base diode is better than the quality of a galena base diode.

So galena base diode is better in its forward characteristic and silicon carbide base diode is better in its' reverse characteristic.

From the experiment data in Table 2, silicon carbide base diode's reserve breakdown voltage is higher than galena base diode's reserve breakdown voltage, so silicon carbide base diode is more stable when the input signal peak-to-peak voltage is large. The reason that causes this

Table 2. Diodes with galena base and silicon carbide base, Fe wire.

Crystal Material	PbS	SiC
Reverse breakdown voltage/V	12.4	>24.9
Inverse current/mA	$\leqslant 1$	$\leqslant 0.006$
Max average forward rectified current/mA	5	1

phenomenon is that for semiconductors, the reverse breakdown voltage increases with the increase of the energy gap. The energy gaps for galena and silicon carbide in different crystal forms of are listed in Table 3.[6,7]

Table 3. Energy gaps of different semiconductors and different crystal forms.

Semiconductor	Crystal form	Energy gap
SiC	$6H\,(> 2200°C)$	3.00
	$15R\,(> 2100°C)$	
	$4H\,(> 1600°C)$	3.25
	$3C\,(< 1600°C)$	2.30
PbS		0.41

Silicon carbide can form four different types of crystal in different temperatures. That's why the crystal of silicon carbide is irregular. All forms of the silicon carbide crystal's E_g value are higher than galena crystal's E_g. Hence, the silicon carbide base diode's reserve breakdown voltage is higher than galena base diode's reserve breakdown voltage.

Interestingly, we also observe weak blue light at the contact point of the silicon carbide base cat's whisker. The formation mechanism is just like LED: electrons in N-region and holes in P-region recombine and produce fluorescence radiation.

To conclude, we compare the properties for galena base and silicon carbide base diodes in Table 4. (under constant conditions of contact quality, contact pressure and material of the thin wire)

Table 4. Properties for galena (PbS) and silicon carbide (SiC) based diodes.

forward resistance	reserve resistance	reverse voltage breakdown	inverse current	max average forward rectified current
PbS<SiC	PbS<SiC	PbS<SiC	PbS>SiC	PbS>SiC

3.6. Cat's Whisker's Application in AM Radio

The most special characteristic of the cat's whisker is its ability to detect AM wave. In this part, we build an AM radio receiver.[8,9]

The function of AM wave is

$$\dot{V}_s = V_s\left[1 + m\cos\left(\Omega t\right)\right]\cos\left(\omega_0 t\right)$$

Here, Ω is the modulation frequency; ω_0 is the carrier wave frequency; m is modulation degree of the AM wave.

Connect the circuit as is shown in Fig. 11.

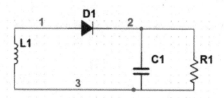

Fig. 11.   AM radio receiver circuit.

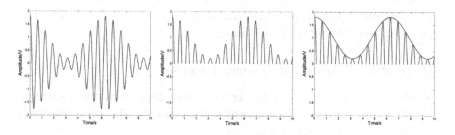

Fig. 12.   Process to get audio signal from AM wave.

The process to get audio signal from AM wave is shown in Fig. 12. First, the AM wave is received by inductance L1. Then, the cat's whisker diode rectifies the current to obtain half-wave signal. Finally, we use the filter capacitor C1 to get the modulation envelope of the input signal.

The diode of the receiver is cat's whisker with galena base and iron wire, which has the best quality according to our experiment data. The capacitor should be air condenser, which is widely used in crystal radios. R1 or the speaker's resistance value should be around 2000 Ω. We wind a copper wire on a square frame ($60cm \times 60cm$) for 20 turns as the inductance. On each turn we left a joint, so the value of the inductance can be changed.

By changing the turns of the inductance and the value of the filter capacitor, we may adjust the AM signal we receive. The music signal we receive by this device is shown in Fig. 13.

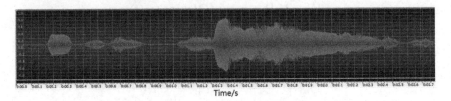

Time/s

Fig. 13.   Music signal received by AM receiver after noise reduction.

4. Discussion

4.1. *Hysteresis Curve*

Increase the forward voltage from zero to its' max forward voltage and decrease the forward voltage to zero to see if the curves will overlap.

As shown in Fig.14 these two curves are very close to each other but not totally overlapped with each other.

4.2. *Diode's I-V Curve Under Extreme Condition*

In order to investigate cat's whisker's electrical properties under extreme conditions, we increase the forward voltage beyond the cat's whisker's working range to destroy it, see Fig.15.

At first the current grows gradually with voltage. When the voltage exceeds a certain value, the current growth becomes irregular.

5. Conclusion

In this article, the main factors to influence the electrical properties are identified as the position of the contact point, the pressure upon the contact point, the material of the wire and the material of the semiconducting crystal.

The quality of cat's whisker decreases with the increase of its' contact pressure. The electrical properties of cat's whisker doesn't change with

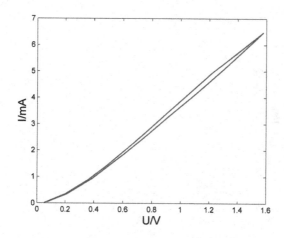

Fig. 14.   Forward voltage from zero to max forward voltage and decrease to zero.

the material of its' metal wire. The properties for galena base and silicon carbide base diodes are compared.

Acknowledgement

I would like to express my gratitude to all those who helped me during the writing of this article.

I would like to express my heartfelt gratitude to Professor Yu Zhang, who led me into the world of physics. I am also greatly indebted to the professors at the School of Physics: Chunfeng Hou and Yuxiao Wang, who have instructed and helped me a lot in the past few months.

Second, my deepest gratitude goes to Professor Sihui Wang, for her guidance on numerous rounds of revision. Without her consistent and illuminating instruction, this article could not have reached its present form.

Last my thanks would go to my beloved family for their loving considerations and great confidence in me all through these years. I also owe my sincere gratitude to my friends and my fellow classmates who gave me their help and time in listening to me and helping me work out my problems during the difficult course of the article.

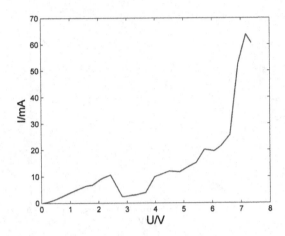

Fig. 15.   Forward voltage higher than diodes working range.

References

1. M. Faraday, *Experimental Researches in Electricity*. Dover Publications (2004).
2. K. F. Braun, On the current conduction in metal sulphides (title translated from german into english), *Ann. Phys. Chem.* (1874).
3. W. Shockley, *Electrons and holes in semiconductors*. D. Van Nostrand Co. (1950).
4. Grundmann and Marius, *The Physics of Semiconductors*. Springer Berlin Heidelberg (2006).
5. H. W. Secor, Radio detector development, *The Electrical Experimenter*. 3, 652–653, 680, 689–690 (1917).
6. K. S. Babu, C. Vijayan, and R. Devanathan, Strong quantum confinement effects in polymer-based pbs nanostructures prepared by ion-exchange method, *Mater. Lett.* **58**, 1223–1226 (2004).
7. F. W. Wise, Lead salt quantum dots: The limit of strong quantum confinement, *Acc. Chem. Res.* **33**(11), 773–780 (2000).
8. A. E. Flowers. Crystal and solid contact rectifiers. http://www.crystalradio.net.
9. Ohm. Build an antique style crystal radio. http://www.instructables.com/id/Build-an-antique-style-crystal-radio.

Chapter 8

2015 Problem 13: Magnetic Pendulum

Chenghao Feng[1]*, Wenkai Fan[1]†, Dachuan Lu[1]‡, Sihui Wang[2], Huijun Zhou[2], Wenli Gao[2]

[1]*Kuang Yaming Honors School, Nanjing University*
[2]*School of Physics, Nanjing University*

Abstract: In this solution, we study the stable oscillations of a magnet pendulum driven by electromagnet connected to an AC power source. The driving force is calculated when considering the electromagnet as a magnetic dipole. The dynamic equation of the system is derived. Based on numerical analysis and experimental observations, the motion of the pendulum has been investigated. Under appropriate driving voltage U, the pendulum can do stable large oscillations when the driving frequencies are odd multiples of the pendulum natural frequency, i.e. $f \approx (2k+1)f_0$. Phase trajectories and phase diagrams are presented. The phase diagram has up to three discrete limit cycles: one for small oscillation, others for big oscillations. The theoretical conditions we found are consistent with our experiment results.

1. Introduction

Problem Statement:

Make a light pendulum with a small magnet at the free end. An adjacent electromagnet connected to an AC power source of a much higher frequency than the natural frequency of the pendulum can lead to undamped oscillations with various amplitudes. Study and explain the phenomenon.

The pendulum in this problem is called "Doubochinski's pendulum",[?] a magnet pendulum is driven by AC powered electric magnet underneath. One characteristic of this pendulum is that the driving force it experiences

---

*E-mail: 141242041@smail.nju.edu.com
†E-mail: 131242013@smail.nju.edu.com
‡E-mail: dclu@smail.nju.edu.com

is both time and coordinate dependent with nonlinear terms. Magnet pendulum is a kind of "kick-excited system" or "self-adapted system"[2] described by the following dynamic equation

$$\ddot{x} + 2\beta\dot{x} + f(x) = \epsilon(x)\Pi(\mu t) \tag{1}$$

For Doubochinski's pendulum, $f(x) = sin(x)$. The magnet driving force is expressed by $\epsilon(x)\Pi(\mu t)$, where $\Pi(\mu t) = cos(\mu t)$.

Mathematical analysis was done with an idealized driving force in the form[2]

$$\epsilon(x) = \begin{cases} F, & |x| \leq d \\ 0, & |x| > d \end{cases} \tag{2}$$

This model with discontinuity in $\epsilon(x)$ in Eq. (2) is obviously oversimplified as compared to the magnet driving force in reality.

Experimental observations discover stable oscillation modes with discrete amplitudes, and explain it from energy perspective or emphasize the effect of driving frequencies.[3,4]

In this paper, we investigate the origin and conditions for this phenomenon. We build a simple experimental setup for demonstration and observation. In our theoretical model, the magnetic driving force will be derived when the electromagnet is regarded as a magnetic dipole. We find that, in nonlinear region, under the same driving frequency and magnitude, the stable oscillation amplitudes have discrete solutions. In addition, the "quantized" oscillations exist only at particular driving frequencies $f \approx (2k+1)f_0$.

The structure of the phase diagram will be examined to reveal the origin of "amplitude quantization". The "quantized" orbits correspond to stable limit cycles. Many of the theoretical conditions we find are verified experimentally.

2. Preliminary experiments

The experimental setup is shown in Fig. 1. We find that when released at small angles, the pendulum does small oscillations at the frequency of the driving force f.[§] Under certain conditions, it does stable large amplitude oscillations at a frequency close to its natural frequency f_0.[¶]

[§]See supplementary materials, magnetic pendulum, video 1.

[¶]See supplementary materials, magnetic pendulum, video 2.

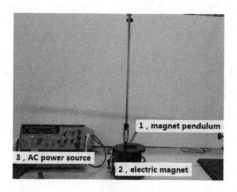

Fig. 1. Experimental setup.

3. The Physics Model

To make theoretical analysis to these phenomena, we write the dynamic equation of the pendulum

$$\ddot{\theta} + 2\beta\dot{\theta} + \frac{mga}{I}sin\theta = \frac{F(\theta)l}{I}cos\omega t \tag{3}$$

where mg is the gravity of the pendulum, a is the distance between the center of gravity of the pendulum and the pivot. I is the rotational inertia of the pendulum. $F(\theta)lcos\omega t$ is the moment of the driving force.

Eq. (3) can be expressed as

$$\ddot{\theta} + 2\beta\dot{\theta} + \omega_0^2 sin\theta = P(\theta)cos\omega t \tag{4}$$

where $\omega_0^2 = \frac{mga}{I}$, $P(\theta) = \frac{F(\theta)l}{I}$.

The pendulum parameters ω_0, β in (Eq. (4)) are obtained by analysing the experimental data when the pendulum is released freely to do damped oscillation. For our magnetic pendulum, $\omega_0 = 5.13$ rad, $\beta=0.08$/s, according to experimental result shown in Fig. 2.

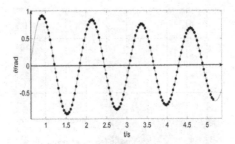

Fig. 2.   θ-t curve of damped oscillation.

The form of driving force $P(\theta)$ in Eq. (4) is essential in determining pendulum's motion.

We regard the magnet as a magnetic dipole whose magnetic moment is M. Then the magnet driving force it experiences is

$$F = \nabla(M \cdot B) \tag{5}$$

where B is the magnetic induction. The electromagnet is also regarded as a magnetic dipole to obtain the magnetic induction. Thus the driving moment $P(\theta)$ the pendulum experiences at any oscillation angle can be defined. Fig. 3 is the numerical result for our experimental setup.

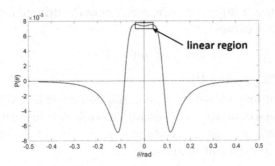

Fig. 3.   The form of driving moment $P(\theta)$ when the electromagnet is regarded as a magnetic dipole.

4. Dynamical Behavior

Because the form of $P(\theta)$ is complex, we can only do theoretical analysis numerically. Firstly of all, we divide $P(\theta)$ into two regions. From Fig. 3, we see that when the oscillation angle θ is small, the driving force $P(\theta)$ is nearly constant. In this region, the pendulum will do forced oscillation. When oscillation angle θ is large, the form of $P(\theta)$ is non-linear. The oscillations in these two areas are different from each other. We will discuss the two situations separately.

4.1. Small Oscillations in Linear Region

When the oscillation angle θ is small, the pendulum only does oscillation in linear region where $P(\theta) \approx P$ is approximately constant, and $sin(\theta) \approx \theta$. So that Eq. (3) can be simplified as

$$\ddot{\theta} + 2\beta\dot{\theta} + \omega_0^2\theta = Pcos\omega t \tag{6}$$

It is exactly the dynamic equation of a forced oscillation. According to Eq. (6), the amplitude of the pendulum satisfies

$$\theta_0 = \frac{P}{\sqrt{(\omega_0^2 - \omega^2)^2 + 4\beta^2\omega^2}} \tag{7}$$

The validity of the approximation in Eq. (6) can be verified by experimental results. Fig. 4(a) shows the experimental result at a fixed driving frequency ($f = 4.08hz$), that the oscillation amplitude is proportional to the driving voltage U, which is proportional to P in Eq. (7). Fig. 4(b) shows amplitude-frequency curve when the driving voltage is fixed. It is a typical resonance curve. The dots are experimental data which fit well with the theoretical curve in Fig. 4(b) obtained from Eq. (7). In this small oscillation region, the oscillating frequency f_p equals the driving frequency f, see Fig. 4(c). In linear region, the theoretical solution matches well with experimental results.

4.2. Large Amplitude Oscillations in Non-linear Region

When the pendulum oscillates in non-linear region, the forced oscillation model will not work any more. Eq. (4) becomes non-linear mainly due to the complicated form of $P(\theta)$, as we see in Fig. 3.

One of the important features of the system is that the pendulum will do oscillations with different amplitudes depending on releasing positions.[3]

To find the conditions when the pendulum can do big stable oscillations, we will conduct experiments as well as simulations by controlling variables such as the driving frequency f, driving voltage U, the initial

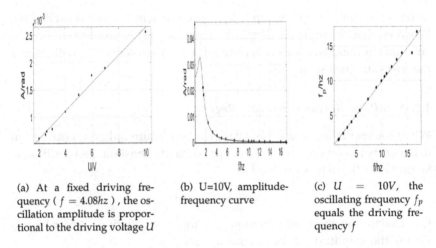

(a) At a fixed driving frequency ($f = 4.08hz$), the oscillation amplitude is proportional to the driving voltage U

(b) U=10V, amplitude-frequency curve

(c) $U = 10V$, the oscillating frequency f_p equals the driving frequency f

Fig. 4.   Experimental results show that small oscillations are forced oscillations.

conditions (here we change the releasing positions θ_0). The effect of these parameters will be discussed separately.

In order to show the roles each variables play in Eq. (4, the dynamic equation can be rewritten as:

$$\ddot{\theta} + 2\beta\dot{\theta} + \omega_0^2 sin\theta = UP_0(\theta)cos(2\pi f t + \phi_0) \qquad (8)$$

4.2.1. *The Effect of Driving Frequency f*

Fix all other variables except the frequency of the AC power source f, we find that the pendulum performs big oscillations only when the driving frequencies are odd multiples of the pendulum natural frequency, $f \approx (2k + 1)f_0$. Otherwise, the pendulum only does small oscillations.

Fig. 5 and Fig. 6 are the phase diagrams according to experimental data. In Fig. 5, $U = 11V$, $f = 2.43hz = 3.01f_0$, the phase diagram contains two discrete phase trajectories. The small circle in the middle corresponds to linear oscillation mode, the large oval corresponds to large amplitude oscillation. In Fig. 6, $U = 11V$, $f = 5.70hz = 5.03f_0$, there are two large amplitude trajectories besides the small one in linear range.

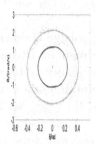

Fig. 5. Experimental phase diagram, U=11V, $f = 2.43hz = 3.01f_0$.

Fig. 6. Experimental phase diagram, U=11V, $f = 4.08hz = 5.03f_0$.

For comparison, we conduct simulations at frequencies $f \approx (2k+1)f_0$. The simulation results are presented in Fig. 7(a), Fig. 7(b), Fig. 7(c) and Fig. 7(d) for $U = 11V$, $f = 3f_0, 5f_0, 7f_0, 9f_0$ respectively. When $f = 3f_0$, there is only one non-linear trajectory. When $f = 5, 7, 9f_0$, there are two non-linear trajectories. When f is too high ($f > 11f_0$), the pendulum will do small oscillations again. These results are all consistent with our experimental observations.

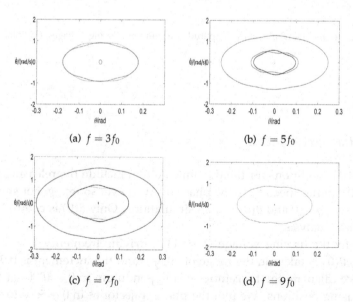

(a) $f = 3f_0$

(b) $f = 5f_0$

(c) $f = 7f_0$

(d) $f = 9f_0$

Fig. 7. Theoretical phase diagrams, $U = 11V$, changing driving frequency.

4.2.2. *The Effect of Driving Voltage U*

The driving voltage U determines the magnitude of driving force. Even if the requirement of driving frequency is satisfied, the pendulum would not do big oscillations if the driving force is small. In experiment, we measured the amplitudes the pendulum can reach at different driving voltages when $f = 5.72hz \approx 7f_0$, see Fig. 8. We find that the pendulum can do large amplitude oscillation when $U > 2.6V$. As U increase further, another large amplitude trajectory appears, see Fig. 9. However, the amplitude increase for a particular trajectory is not sensitive to the driving voltage. See Fig. 8, when U increase for 3-4 times, the amplitude increase of the trajectory is only about 10%.

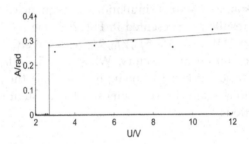

Fig. 8.   Experimental results of the amplitudes at different driving voltages, f=5.70hz $\approx 7f_0$.

4.2.3. *The Effect of Releasing Positions*

The initial conditions for the dynamic system include the releasing position $\theta(0)$, initial speed $\dot{\theta}(0)$, the phase of AC power source ϕ. For simplicity, we set $\phi_0 = 0$ and $\dot{\theta}(0) = 0$ in simulations. Only $\theta(0)$ is controlled in numerical analysis.

We fix the driving voltage $U = 11V$, driving frequency $f = 7f_0 = 5.63hz$. $\theta(0)$ is taken in the range of $(0, \frac{\pi}{2})$, and the increment is $0.02rad$. Then we calculate the trajectories of the pendulum in $0 - 1000s$ at these 76 releasing positions. We plot the phase trajectories in $0 - 500s$ to show transient oscillations, and the $500 - 1000s$ to show stable oscillations.

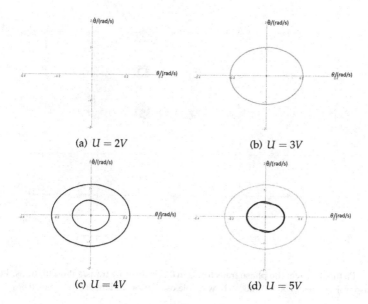

(a) $U = 2V$ (b) $U = 3V$

(c) $U = 4V$ (d) $U = 5V$

Fig. 9. Theoretical phase diagrams, $f = 5f_0$, changing driving force.

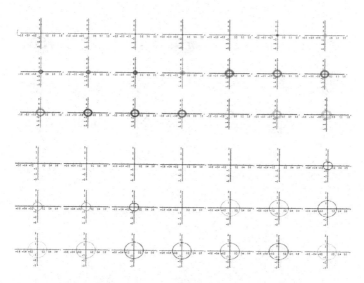

Fig. 10. Pictures above: the phase trajectories in $0 - 500s$ show transient oscillations. Below: the phase trajectories in $500 - 1000s$ show stable oscillations. Here $\theta(0) \in (0, 0.42)$.

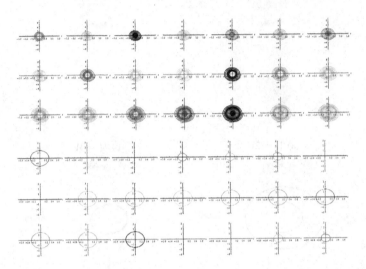

Fig. 11.   Pictures above: the phase trajectories in $0 - 500s$ show transient oscillations. Below: the phase trajectories in $500 - 1000s$ show stable oscillations. Here $\theta(0) \in (0.44, 0.86)$.

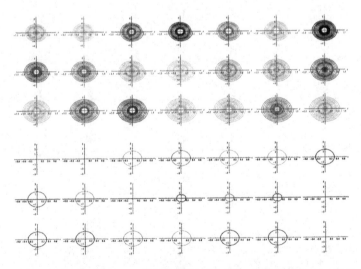

Fig. 12.   Pictures above: the phase: trajectories in $0 - 500s$ show transient oscillations. Below: the phase trajectories in $500 - 1000s$ show stable oscillations. Here $\theta(0) \in (0.88, 1.30)$.

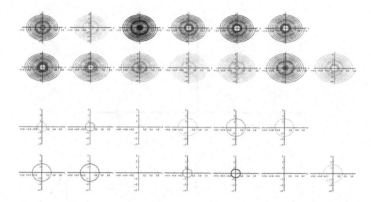

Fig. 13. Pictures above: the phase trajectories in 0 − 500s shows transient oscillations. Below: the phase trajectories in 500 − 1000s show stable oscillations. Here $\theta(0) \in (1.32, 1.50)$.

Fig. 10,11,12 and 13 show the simulation results for $\theta(0) \in (0, 0.42)$, $\theta(0) \in (0.44, 0.86)$, $\theta(0) \in (0.88, 1.30)$ and $\theta(0) \in (1.32, 1.50)$ respectively. We find that the pendulum released at small initial angle $\theta(0)$ will only do small oscillations. When $\theta(0)$ is relatively big, say > 0.24 rad, it might either do large oscillation in non-linear region or small oscillations in linear region. It might be suspected that different $\theta(0)$ will lead to numerous amplitudes of oscillations. However, when we put all the 76 phase diagrams into one picture, we find that all the stable trajectories collapse into discrete cycles of three different amplitudes as shown in Fig. 14.

According to above experimental and simulation results, the pendulum has one small oscillation and up to two big oscillation amplitudes at given driving frequency f and voltage U. Among these limit cycles, whichever a pendulum will finally evolve into depends on initial conditions.

5. Discussions

5.1. *What If We Suppose the Driving Force is a "Square Wave" Driving Force as the Reference Does?*

For comparison, we also simulated the the results for "square wave" driving force, see Fig. 15. We take the form of the "square wave" $P(\theta)$ driving force as

$$P(\theta) = \begin{cases} P, & |\theta| \leq d \\ 0, & |\theta| > d \end{cases} \tag{9}$$

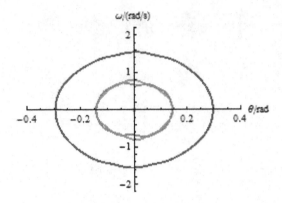

Fig. 14.   In the phase diagrams of different initial conditions, all stable trajectories collapse into discrete cycles of three different amplitudes.

and $d = 0.01, P = P_0$.

In Fig. 15, large oscillation modes are also found at $f = (2k + 1)f_0$. However, they do not fit the experimental results satisfactorily. At low frequencies, it fails to describe the complicated behaviors qualitatively. At higher frequencies, the amplitudes of the oscillation modes are inconsistent with experiment results.

(a) $f = 3f_0$ (b) $f = 5f_0$ (c) $f = 7f_0$ (d) $f = 9f_0$ (e) $f = 11f_0$

Fig. 15.   Simulation results of phase diagrams, "square-wave" driving force.

6. Conclusion

We build a simple experimental setup for demonstration and observation. Theoretical model has been derived based on dipole model of the electromagnetic driving force. According to our experimental and simulation results, Pendulum released at small initial angle will only do small oscillations. Under appropriate driving voltage U, the pendulum can do stable large oscillations at driving frequencies $f \approx (2k + 1)f_0$. At given driving frequency f and voltage U, the phase diagram has up to three discrete amplitudes: one for small oscillation, and two for big oscillations. Among these oscillation modes, whichever a pendulum will finally evolve into depends on initial conditions.

References

1. D. Penner, Y. B. Duboshinskii, D. Duboshinskii, and M. Kozakov. Oscillations with self-regulating interaction time. In *Soviet Physics Doklady*, vol. 17, p. 541 (1972).
2. V. Damgov and I. Popov, discrete oscillations and multiple attractors in kick-excited systems, *Discrete Dynamics in Nature and Society*. 4(2), 99–124 (2000).
3. J. Tennenbaum, Amplitude quantization as an elementary property of macro-scopic vibrating systems, *21ST CENTURY SCIENCE AND TECHNOLOGY*. **18** (4), 50 (2006).
4. D. Doubochinski and J. Tennenbaum, On the general nature of physical objects and their interactions, as suggested by the properties of argumentally-coupled oscillating systems, *arXiv preprint arXiv:0808.1205* (2008).

Chapter 9

2015 Problem 14: Circle of Light

Xiaodong Yu[1]*, Boyuan Tao[2]†, Xin Yuan[3]‡, Sihui Wang[1], Wenli Gao[1]

[1]*School Of Physics, Nanjing University*
[2]*Kuang Yaming Honors School, Nanjing University*
[3]*School of Chemistry and Chemical Engineering, Nanjing University*

A circle of light has been observed experimentally. With the increase of the distance between the incident point and the wall, different pattern and interference structure can also appear on the circle of light under proper conditions. Using the theory of geometrical and wave optics, we give an explanation to figure out the formation of the circle and its structures. Besides, some discussions on the calculations of the intensity distribution have been made under different circumstances.

1. Introduction

When a laser beam is aimed at a wire, a circle of light can be observed on a screen perpendicular to the wire. Explain this phenomenon and investigate how it depends on the relevant parameters.

This problem asks us to explore what the phenomenon is like and what influences the circle's shape, size, pattern etc. Many reserchers have already studied part of this problem. H.J.Kong and Jin Chol[1] from Korea Advanced Institute of Science and Technology have studied the formation of the circle but they didn't mention the phenomena of diffraction and interference. J.Segal, A.Cedarman and D.Maclsaac from Buffalo State University have observed the diffraction with hair or thin wire. S.Ganci[2] has published an article about "Fraunhofer diffraction by a thin wire and Babinet's principle". We would like to use a different method to figure out

*E-mail: yuxiaodongnju@gmail.com
†E-mail: 15905172626@163.com
‡E-mail: 141130130@smail.nju.edu.cn

the formation and intensity distribution of the circle as well as diffraction and interference phenomena.

In the following sections, we will first reproduce the phenomenon to find the differences between different circles of light. Then we will introduce a straightforward explanation to make it clear what affects the circle's shape. Factors like diameters of wires, angles between the wire and the screen are considered in this explanation. Next, we are going to calculate the light intensity when the pattern is a perfect circle. During the experiments, we found that there exist diffraction and interference patterns in the circle of light which we will also give some interpretations and calculations. Finally, we will make a summary on what we have studied.

2. Experimental Arrangement

The set up used in the experiment is shown in Fig. 1(a). The laser beam is aimed at the fixed wire. The wire is perpendicular to a wall which is used as a screen. The diameters of the wire and the laser beam are $0.80mm$ and $1.0mm$ respectively. A sketch of the set-up is given in Fig. 1(b) to show the relationships among the laser beam, the wire and the screen clearly.

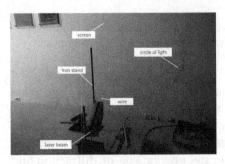
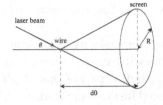

(a) Experimental set-up to the problem. (b) A sketch to the set-up, where R is the radius of the light circle and d_0 is the distance between the incident point and the wall.

Fig. 1.   Experimental installation of the circle of the light.

Intensity of the light on the circle is measured by optical power meter. The pictures are photographed by Pioneer L2035AW.

3. Phenomena and Explanations

Keeping the angle between the laser beam and the wire constant, and changing the angle between the wire and the screen, different shapes with different sizes and light intensities will be observed on the screen. We will discuss these phenomena from the perspective of geometric optics in this section.

3.1. *Shape and Size*

Fig. 2 shows clearly that with the decreasing of the angle between the wire and the screen, changing from $\pi/2$ to 0, there appear different shapes of light including a perfect circle, ellipse, parabola and hyperbola on the screen, as showing in Fig. 3.

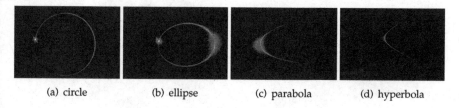

(a) circle   (b) ellipse   (c) parabola   (d) hyperbola

Fig. 2. Different curves on the screen.

Obviously, all the light shapes go through a series of conic section. Next, a geometrical model would be built to figure out how the angle influences the shape of light.

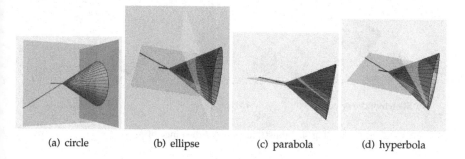

(a) circle   (b) ellipse   (c) parabola   (d) hyperbola

Fig. 3. Simulations of different curves.

Based on the experimental conditions, we assume that the wire is a glossy cylinder with the diameter less than that of the laser. So, the dominant reflection here is specular reflection rather than the diffuse one.

We begin with a projection to the transverse plane. Watch the wire axially so that all the light rays and the surface normal are projected onto the transverse plane of the wire. We define ϕ as the incident angle on the projection.

For a thin wire, as the diameter of the laser beam is larger than that of the wire (as shown in Fig. 4(c)), the laser beam can cover the cross section of the wire, insuring ϕ to take any value from 0 to 90 degree (Fig. 4(d) and Fig. 4(e)).

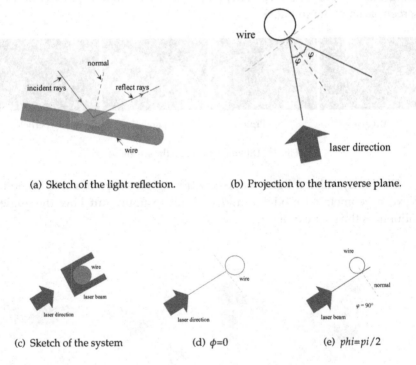

(a) Sketch of the light reflection.   (b) Projection to the transverse plane.

(c) Sketch of the system   (d) $\phi=0$ (e) *phi=pi/2*

Fig. 4.   Illustrations about the range of ϕ.

Now, we take a lateral view. It is easy to get from reflection law that the angle between the reflected ray and wire is equal to that between the incident ray and the wire. Under the precondition that the incident laser rays are parallel, they all have the same angle to the wire. So the reflected rays have the same bias angle to the wire which is defined as θ (Fig. 5(a)). Since the distance from the wire to the screen is much larger than the diameter of the wire, all the reflected rays are considered from the same point. As a result, they have the same angle with respect to the wire. With consideration to these two facts, we can conclude that all the reflected rays are on the same conic surface and the wire is the axis. (Fig. 5(b))

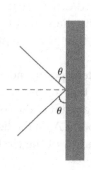

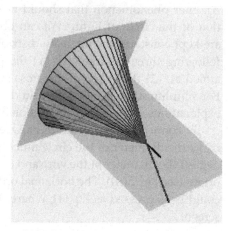

(a) Literal view of the reflection.   (b) Reflected lights are on a conic surface.

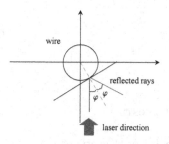

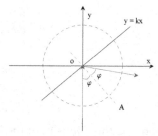

(d) The red circle is the light circle observed on the screen

Based on the above analysis, it can be known that the shape of light on the screen depends on the angle between the wire and the screen, and that if the wire is perpendicular to the screen, a perfect circle can be observed. As to the radius of the circle, it depends on at least two relevant parameters. One is the angle θ between the incident rays and the wire and the other one is the distance between the incident point and the screen. Qualitatively, the larger the angle is or the larger the distance is, the larger the radius is.

3.2. *Intensity Distribution*

Another phenomenon that should be discussed is the intensity distribution of the circle of light. We can observe that when other parameters are kept constant, the intensity distribution on the circle is influenced by following three parameters: (1) the position on the wire that the laser aimed at, (2) the shape of the cross section (3) the diameter of the wire. For simplification, we will only study the case of thin glossy wire. Other explanations will be given in Section 5.

By controlling the distance between the incident point and screen to make the diameter of the circle much larger than that of the wire, we can neglect the diameter of the wire and consider it as a point at the coordinate origin (see Fig. 5(d)). The horizonal ordinate x at point A on the light circle could be expressed as Eq. (1), where, R_w is the radius of the circle on the screen.

$$x = R_w \sin \phi \qquad (1)$$

Then

$$\frac{dx}{d\phi} = R_w \cos \phi \qquad (2)$$

As

$$\frac{dI}{d\phi} = \frac{dI}{dx}\frac{dx}{d\phi} = \frac{dI}{dx}R_w \cos \phi \qquad (3)$$

Where I is the intensity of the light at point A. We can get

$$\frac{dI}{dx} = \frac{\csc \phi}{R_w}\frac{dI}{d\phi} = P\csc \phi R_w \qquad (4)$$

Then we design an experiment to verify our theory. As we translate the wire in the direction perpendicular to the laser beam, the peak of the transverse distribution of the laser is also translated, because as shown in the diagram, the x position of the laser beam peak intensity on the wire surface is changed. x is replaced by x plus Δx, where Δx is the translation length.

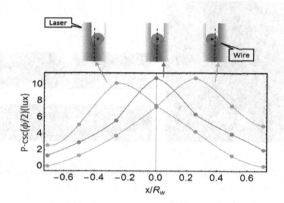

From Eq. (4), we know that three factors mainly affect the intensity distribution on the circle, which are the position that the laser aimed shape of the wire and the diameter of the wire. Shape of the wire and the diameter of the wire affect the intensity distribution by affecting the derivative of ϕ with respect to x.

The intensity transverse distribution of the laser is determined by the laser device itself and the position it aims at the wire surface. Considering the laser beam is a Gaussian beam, the intensity distribution should obey the Gaussian distribution. Here we transform the experimental data of the intensity angular distribution to the intensity transverse distribution of the laser, and find the graph is similar to the Gaussian distribution.

3.3. Completeness of the Circle

In our experiments, we find that there exist some parameters which will affect the completeness of the circle. Eq. (4) tells that the position the laser aimed at, the shape of the wire and the diameter of the wire are three main factors that influence the completeness of the circle.

3.4. Diffraction Pattern

Another intriguing phenomenon observed is that with a thin enough wire and a proper distance between the incident point and the screen, some fringes appear around the brightest dot on the light circle (see Fig. 5), which can be thought to be the effect of diffraction.

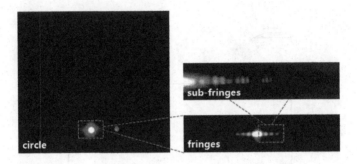

Fig. 5.   In a perfect circle, there are secondary and tertiary structures.

3.4.1. *Diffraction*

The optical path for diffraction is shown in Fig. 6.

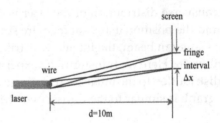

Fig. 6. a is the diameter of the wire, Δx is the interval of the fringes, d_0 is distance from the incident point on the wire to the brightest dot on the circle between the laser beam and the wire.

According to the Babinet's principle, the fringes interval Δx here can be got from that of the single slit. So we use the formula from the single slit.

$$\Delta x = \frac{\lambda d}{a} = \frac{\lambda d_0}{a \cos \theta} \tag{5}$$

Using the parameters showing in Fig. 6, we can calculate the theoretical fringes interval Δx to be $1.24cm$. The measured value of Δx from the diffraction pattern is about $1.21cm$, which is in good agreement with the theoretical one.

3.4.2. *Interference*

When we further change the distance d to a larger extent, we find that there appear sub-fringes in each bright diffraction fringe of the circle (Fig.5), which could be attributed to the interference between the refraction rays and the reflected rays. Fig. 6 is the diagram of the optical path of the reflected light and diffraction light are shown.

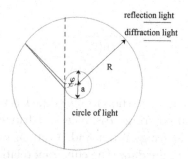

Fig. 7. Interference of the diffraction rays and the reflected rays causes the sub-fringes in the diffraction fringes.

Introduces Taylor series expansion. Due to small-angle approximation, optical path difference is derived to be

$$\Delta = \frac{\sqrt{aR}}{32}\phi^3 \qquad (6)$$

where $\phi = \dfrac{x}{R}$

The optical path difference is in proportion to cube of ϕ. When the optical path difference is odd multiple of half wavelength, we get the dark fringe position x in the fringe structure, which is in proportion to the cube root of fringe order.

When $\Delta = 2k + 1\dfrac{\lambda}{2}$, there is

$$x = \sqrt[3]{\frac{16(2k+1)\lambda}{\sqrt{aR}}R^3} \qquad (7)$$

x of dark fringe is in proportion to $\sqrt[3]{k}$ while k is the fringe order.

We see the measured fringe distance and fringe order. Due to the uncertain of the initial fringe order and position, we use an expression with three various to fit the data. The cube root relation is confirmed.

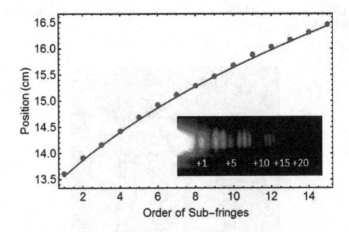

Fig. 8.   Cube root relation.

4. Conclusion

In this work, we observed four kinds of 'Circles of Light' and meanwhile each light circle has three levels of the structure. Different angles between the screen and the reflected lights result in different types of "Circles of Light'. To explore the reliance between the circle's shape and the angle, we built a reflection model. If the wire is perpendicular to the screen, it is a circle. If the angle is smaller than the angle of cone, it is an ellipse. If equals to the angle of cone, it is a parabola and it is a hyperbola while the angle is larger than the angle of cone. The distances between the incident point with the screen lead to various sizes. In the situation of a critical circle, we also calculated the intensity distribution. As to three levels of the structure, we use diffraction and interference theories to explain that. In the interference part, we give an equation describing the relationship between fringe position and fringe order.

5. Discussion

(1)Q: If the wire diameter is changed, what will happen?

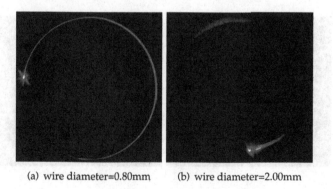

(a) wire diameter=0.80mm   (b) wire diameter=2.00mm

Fig. 9.   The affect of the diameters of the wire.

A: In these experiments, we also changed some other parameters and we are happy to find that the diameter of wire and the position that the laser is aimed at will influence the completeness of the circle. According to the foregoing model, the change of the wire diameter will affect the shape of light and completeness of the circle. However, if the wire surface is not glossy as assumed, some details will be different. Like shown in the picture, the second circle of light seems to be thicker. The wire surface is not glossy, which is attributed to the diffuse reflect by the rough surface. The thicker wire lead to more diffuse reflection area, inducing a dim but wide circle of light.

(2) Q: What's the influence of the shape of wire caused on the circle of light?

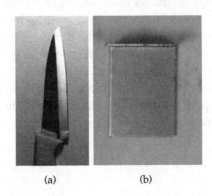

(a)   (b)

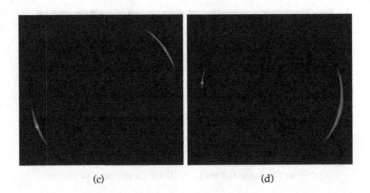

(c)   (d)

A: We tried different thin transverse section used as wire. The shapes of the circle are almost the same while the light intensity distribution are obviously different.

(3) Q: What happens if the wire is not glossy enough?
A: The circle will look dimmer on account of the change of the angle between the reflected rays and wire.

(4) Q: What happens if the laser beam is not collimated enough?

A: Light rays in the uncollimated laser beam have different bias angle, so the reflected light will not have a uniform angle with respect to the wire and results in a dim circle on the screen.

(5) Q: What happens considering the grains on the wire surface?

A: Axial grains change the shape of transverse section, so they influence the intensity distribution but not the shape of the circle of light. Cross grains make the wire no longer straight at the incident point. They will cause aberrance on the circle of light which is also called scattering.//

References

1. H. J. Kong, J. Choi, J. Shin, S. W. Yi, and B. Jeon, Hollow conic beam generator using a cylindrical rod and its performances, *Optical Engineering*. 45(8), 084005–084005 (2006).
2. S. Ganci, Fourier diffraction through a tilted slit, *European Journal of Physics*. 2 (3), 158 (1981). URL http://stacks.iop.org/0143-0807/2/i=3/a=006.

Chapter 10

2015 Problem 15: Moving Brush

Yiran Deng[1]*, Boyuan Tao[2]†, Bingnan Liu[2], Sihui Wang[3], Huijun Zhou[3]

[1] *School of Earth Science and Engineering, Nanjing University*
[2] *Kuang Yaming Honors School, Nanjing University*
[3] *School of Physics, Nanjing University*

A brush may begin translational and rotational motion when placed on a vibrating horizontal surface. In this paper, we analysis the causes of the movement, the trajectories, and how the parameters such as the vibrating amplitude and frequency influence the motion of the brush. With adequate understanding of the rule behind its motion, we manage to control a brush's motion effectively.

1. Introduction

Problem Statement:

A brush may start moving when placed on a vibrating horizontal surface. Investigate the motion.

A popular DIY toy robot called "Bristlebots"[1] uses conventional toothbrush head and micro-motors to realize locomotion. Similarly, a brush may start moving when placed on a vibrating horizontal surface. The effect of oscillatory contact on sliding motion has been studied concerning the dynamic model of sliding motion,[2] the influence of perpendicular and tangential oscillation contact to the magnitude of friction,[3] and relation between ultrasonic oscillation and sliding friction.[4] Gutowski, P. et al.[5] have found that oscillation decreases the sliding friction and gives the threshold of occurrence. It is supposed that apart from the amplitude and frequency, stiffness and stress count, too. The effect of anisotropism to the vibratory machine has been proposed, investigated in applications in

---

*E-mail: dengyiran_nju@163.com
†E-mail: 15905172626@163.com

robot engineering and biology.[6-8] In this paper, we attribute the causes of the movement to the anisotropic friction due to oscillatory contact. We will investigate the trajectories of brushes under different conditions to find the mechanism.

By controlling parameters, we find how vibrating amplitude and frequency influence the motion of the brush. With adequate understanding of the rule behind its motion, we manage to control a brush's motion effectively.

2. Preliminary Experiment

First of all, we conduct simple experiments to reproduce the phenomenon and find typical modes of brushes' motion. The experimental apparatus includes a vibration platform (0-100Hz, 500W), and a variety of brushes, see Fig. 1. The vibration platform produces vertical vibration with adjustable frequencies and amplitudes.

(a) Vibration platform   (b) A variety of brushes

Fig. 1.   Experimental apparatus.

In the first preliminary experiment, we put different types of brushes (a rectangular plastic brush, a rectangular mane brush, a roun mane brush) on the vibration platform and observe their motion, see Fig. 2a, b, c [‡]. We find that they all have translational and rotational motion. The velocities and directions of rotation of the brushes are independent of their shapes, but mane brushes' velocities are greater than plastic brushes'.

[‡]See supplementary materials, moving brush, video 1, 2, 3

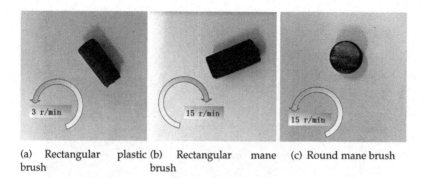

(a) Rectangular plastic brush (b) Rectangular mane brush (c) Round mane brush

Fig. 2. Motion of different types of brush.

In the second experiment, we put two round mane brushes on the platform in turn. We still find that they have different ways of motion: one goes clockwise, the other goes counter-clockwise. We suspect that the condition of bristles is a controlling factor to its movement, see Fig. 3a, 3b[§].

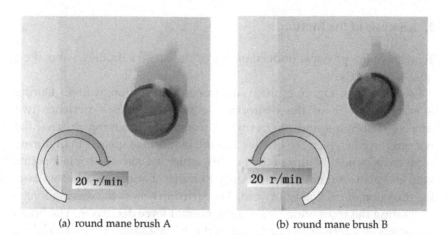

(a) round mane brush A   (b) round mane brush B

Fig. 3. Two brushes of the same type have different ways of motion.

Finally, we use brushes with rigid bristles, see Fig. 4. They only perform slight translational motion without rotation at all! This experiment shows the importance of the bristles in determining the brush's motion.

[§]see supplementary materials, moving brush, video 4, 5

Fig. 4.   Brush with rigid bristles only performs slight translational motion without rotation.

3. Analysis of the Motion

To achieve the physical understanding of the brush's motion, we make a qualitative analysis.

As shown in Fig. 5, a brush undergoes acceleration at first. During a vibrating cycle of the platform, the brush bristles experience two process–deforming and recovering. During the deforming process, the normal force between the bristles and the platform is greater than during recovering, while the tips of the bristles, which are also the contact points with the platform, have the tendency to move "backward". In this case, the direction of friction is forward. Thus friction acts as the driving force.

During the recovering process, the normal force decreases. The brush has acquired a forward velocity due to the driving force. There is a sliding between bristles and platform. So that friction acts as the resistance in this case. But the resistance is smaller than the driving force because of the smaller normal force. Thus the brush will get accelerated.

Once the brush acquires a forward velocity, even in the deforming process, the contact points may move forward. So friction may act as resistance for a longer period of time. Resistance will increase when the brush picks up velocity. Uniform motion can be reached when the driving force is balanced by the resistance.

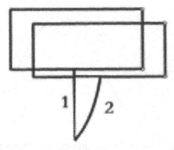 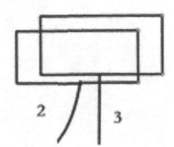

(a) Process of deforming.   During deforming, friction mainly acts as the driving force.

(b) Process of recovering. During recovering, friction mainly acts as the resistance.

Fig. 5.   Deforming and recovering of brush bristles.

Note that the net force acted on the brush is the vector summation of all the forces on each of the bristles. As we use the resulting force to simplify the brush motion, the force might not be located right at its center of mass. Therefore, it produces a net torque with respect to the center of mass as well as the net force. This explains why the brush's translational motion is usually accompanied by rotation.

4. Trajectories

In order to analyze the brushes' motion, we use high-speed camera to record their motion and use "Tracker" to acquire data by analyze the video clips.

Typically there are two kinds of trajectories for moving brushes: circular and cycloid.

When the platform is horizontal, it is circular, see Fig. 6. In experiment, we mark two points on the brush. Fig. 6 shows that the tracks of these points are concentric circles.   This means that the rotational angular velocity is the same as the angular velocity of the center of mass. We can describe the circular motion with angular velocity and radius of the orbit R. The data of x/y coordinates fit with the displacement time curve of a uniform circular motion, as shown in Fig. 6b.

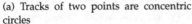

(a) Tracks of two points are concentric circles

(b) Displacement-time curve of uniform circular motion

Fig. 6.   A brush performs circular motion.

When the platform is not perfectly horizontal, the orbit will become a cycloid. Fig. 7a shows a cycloid trajectory in experiment. In Fig. 7b shows the experimental data and the displacement time curve of a cycloid.

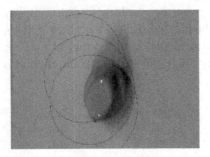
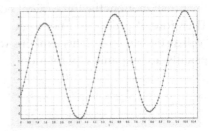

(a) Trajectory of cycloid

(b) Experimental data and displacement-time curve of cycloid

Fig. 7.   A brush performs cycloid motion.

5. Controlling Parameters

5.1. *The Effect of Frequencies*

Frequency and amplitude are most important parameters to describe a vibration. In this section, we analyze their influence on the brush motion separately.

First of all, we measure the brush's angular velocities ω under different frequencies f of the vibration platform. The result is shown in Fig. 8. The different driving amplitudes are indicated with colors (red, blue, grey from big to small).

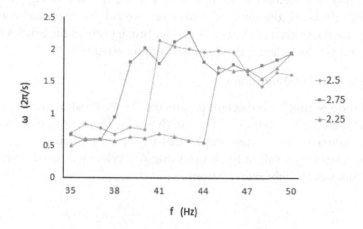

Fig. 8. The brushs angular velocity-frequency curves at different driving amplitudes. The different driving amplitudes are indicated with colors (red, blue, grey from big to small).

We find that in the diagram, there are "jumps", or sudden increase in angular velocities, at some frequencies, and "jumps" occur at smaller frequencies for larger amplitudes.

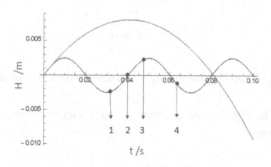

Fig. 9. h-t diagram, in phase motion between the platform and the brush.

We tentatively explain the "jump" as "in phase" contact between the platform and the brush. Fig. 9 is a height -time diagram. The blue curve is for the vibration platform, the orange one is for the brush. As the brush falls and making contact with the platform, the platform can be rising (2), falling (4) or nearly still (1,3). In each situation, the impact the brush experiences is different. If the platform is rising (point 2 in Fig. 9) when a brush falls on it, the impact is stronger. We call it "in phase" contact. Otherwise the impact is weaker. A greater impact causes the bristles bend more and produces larger driving force and forward speed.

5.2. *The Effect of Amplitudes*

Amplitude is another important parameter. As we illustrated, "in phase" contact may cause "jump". So we choose a frequency to avoid "in phase" contact and change amplitudes to see its influence. Fig. 10 is the experimental result of the brush's angular velocity ω under different amplitudes of the vibration platform.

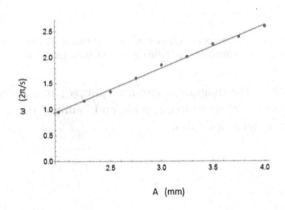

Fig. 10.   Brushs angular velocity under different driving amplitudes.

As shown in Fig. 10, the angular velocity increases linearly with the driving amplitude.

5.3. *Orbit Radius r is Constant!*

In the experiment, we discovered an interesting phenomenon. For the same brush under different conditions, the orbit radius is a constant! Change the frequency while fix other parameters, the radius is nearly the same, see Fig. 11.

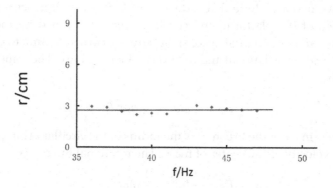

Fig. 11. Brushs radius under different frequency is a constant.

Then, we change the roughness of the platform surface. We stick a piece of sand paper on the table, one with rough surface up, the other with smooth surface (back side) up, see Fig. 12.[¶] We find that the velocity of the brush on the rough surface is smaller. But it is noteworthy that the radius of the trajectories of the motion are still the same!

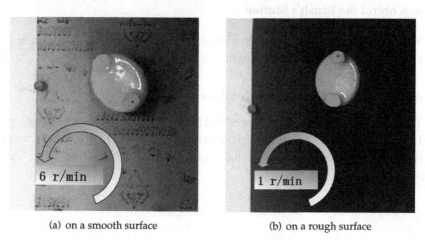

(a) on a smooth surface (b) on a rough surface

Fig. 12. Changing the roughness of the surface, the radius of are still the same.

[¶]See supplementary materials, moving brush, video 6, 7.

So, we may conclude that radius is irrelevant to frequencies and the roughness of the platform surface. It is only determined by the brush itself. We can also infer that a constant radius is corresponding to constant ratio between translational and rotational kinetic energy. The translational energy is

$$E_T = \frac{1}{2}mv^2 = \frac{1}{2}m\omega_1^2 R^2$$

where m is mass of the brush, R is the radius of the circular orbit. ω_1 is the angular velocity of the center of mass. The rotational Energy is

$$E_R = \frac{1}{2}I\omega_2^2 = \frac{1}{2}km\omega_2^2 r_0^2$$

Where r_0 is radius of the brush, ω_2 is the rotational angular velocity, which equals ω_1. I is the inertia of moment of the brush.

If the ratio of radius is constant $a = \frac{R}{r_0}$, then we have:

$$\frac{E_T}{E_R} = \frac{1}{k}(\frac{R}{r_0})^2 = \frac{1}{k}a^2$$

6. Control the Brush's Motion

With above understanding of the brush's motion, we try to apply them to control the movement of the brush.

6.1. *Control the Direction of Rotation*

The direction of rotation is determined by the bristles state of deformation. In experiment, we wring the bristles along the orientation opposite to its natural state. Then we observe that brush moves in a "negative" direction, see Fig. 13.

Fig. 13. Control the direction of rotation by changing the bristles' orientation.

6.2. Control the Velocity

When the resistance equals the driving force, it will reach a stable motion. By reducing the coefficient of friction of the vibrating surface, the brush can reach a higher equilibrium velocity. We can also control the velocity by adjusting the driving amplitudes and frequencies.

Fig. 14. Change the length distribution of bristles.

6.3. Control the Orbit Radius

We change the length distribution of the bristles by cutting half of them short, see Fig. 14. Anisotropism is enhanced due to the uneven length distribution. When we repeat the experiment, we find that the amended brush has a smaller radius, see Fig. 15.

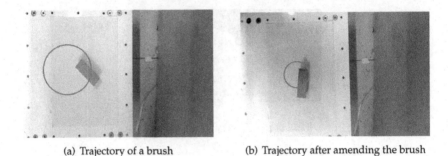

(a) Trajectory of a brush   (b) Trajectory after amending the brush

Fig. 15.   Control the orbit radius by changing the the bristles length distribution.

7. Conclusion

A brush may begin translational and rotational motion when placed on a vibrating horizontal surface. In this paper, we analysis the causes of the movement, the trajectories, and how the parameters such as the vibrating amplitude and frequency influence the motion of the brush. The trajectory of the motion is a circle when the platform is horizontal. The trajectory becomes a cycloid when the platform is not perfectly horizontal. Rotational angular velocity is the same as the angular velocity of the center of mass. The relation between angular velocity and frequency is not monotonic. We explain the "jump" as "in phase" contact between the platform and the brush. The angular velocity increases linearly with the driving amplitude.

We discover that for the same brush, the orbit radius is a constant under different conditions, indicating that the ratio of translational energy and rotational energy is constant.

With adequate understanding of the rule behind its motion, we manage to control a brush's motion effectively. We control the direction of rotation by changing the orientations of the bristles, control the angular velocity by changing the roughness of the vibrating surface, control the orbit radius by changing the length distribution of the bristles.

References

1. Educational Robotics Kits Bristlebots.LINK: http://www.bristlebots.org/
2. Bengisu M. T, Akay A. (1999). StickCslip oscillations: Dynamics of friction and

surface roughness. The Journal of the Acoustical Society of America,105(1), 194-205.

3. Leus M., Gutowski P. (2008). Analysis of longitudinal tangential contact vibration effect on friction force using Coulomb and Dahl models. Journal of Theoretical and Applied Mechanics, 46(1), 171-184.

4. Littmann W., Storck H., Wallaschek J. (2001). Sliding friction in the presence of ultrasonic oscillations: superposition of longitudinal oscillations.Archive of Applied Mechanics, 71(8), 549-554.

5. utowski P., Leus, M. (2012). The effect of longitudinal tangential vibrations on friction and driving forces in sliding motion. Tribology International, 55, 108-118.

6. Lysenko V., Zimmermann K., Chigarev A., Becker F. (2011). A mobile vibro-robot for locomotion through pipelines. Technology, 12, 16.

7. Morin J. W. (2011). Design, fabrication and mechanical optimization of multi-scale anisotropic feet for terrestrial locomotion (Doctoral dissertation, Massachusetts Institute of Technology).

8. Filippov A., Gorb S. N. (2013). Frictional-anisotropy-based systems in biology: structural diversity and numerical model. Scientific reports, 3.

Chapter 11

2015 Problem 17: Coffee Cup

Dachuan Lu[1]*, Wenli Gao[2], Huijun Zhou[2]

[1]*Kuang Yaming Honors School, Nanjing Univerity*
[2]*School of Physics, Nanjing University*

1. Introduction

Physicists like drinking coffee, however walking between laboratories with a cup of coffee can be problematic. Investigate how the shape of the cup, speed of walking and other parameters affect the likelihood of coffee being spilt while walking.

This problem has aroused many researchers' interests for many years, Wedemeyer and Reese first attached to the problem of liquid pitching oscillations in an upright circular cylindrical cup in 1953.[1]   And in 1994, Takahara investigated the frequency response of pitching vibration.[2] Other types of vibration are investigated experimentally by Brown in 1954.[3] In our study, simplified walking model is extracted from reality situation to analyze the excitations to the cup, and fluid mechanics is used to solve the coffee motion in the cup. From the analysis, we can find how the parameters affect the likelihood of the coffee being spilt.

Normally, there are two ways of coffee being spilt out, namely, splashing and spilling. Splashing is defined as a phenomenon that liquid layers collide the cup wall, leading to the drops splashing out. Spilling appears when the frequency of excitation is close to the natural frequency of the water motion, causing resonance, which will lead to liquid level exceeding the cup wall. Generally, the frequency of human excitation ranges from 1hz to 3.5hz, which is close to the natural frequency of the coffee in the cup and will mainly cause it to spill out. Our study will mainly focus on the mechanism of spill.

---
*E-mail: dclu@smail.nju.edu.com

In this work, we classify the human motion into 3 stages as shown in Fig. 1, namely, acceleration, steady motion and deceleration stage, suggesting that various excitations are exerted to the cup in each stage. To be specific, in the 1st stage, the coffee undergoes a uniform acceleration process leading the liquid surface to be slope. In the 2nd stage, periodical excitations, such as horizontal, vertical and pitching vibration dominate. In the 3rd stage, a uniform deceleration excitation is exerted to the cup. After human motion stops, the coffee in the cup will still experience the fourth stage: a free vibration.

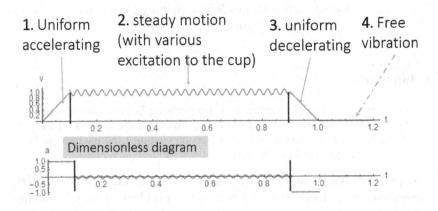

Fig. 1.   Classification of human motion.

In this article, theoretical model is developed to analyze the different stages in the simplified walking model, while corresponding experiments have been conducted to verify them.

2. Model Development

2.1. *Assumptions*

As to this problem, the following assumptions can be given

(1) As $\frac{1}{Bo} = \frac{\sigma}{\rho g R^2} \sim 10^{-2}$, the surface tension can be neglected.
(2) As the coefficient of compressibility β is about $10^{-10} Pa^{-1}$, the density can be regarded as a constant.

(3) For the degree of damping $\Lambda = \frac{\beta_\omega}{\omega_0} \sim 10^{-2}$ is two order less than 1, viscosity can be neglected.

(4) The fluid is assumed to be irrotational flow.

where σ is the surface tension coefficient, ρ is the density of the water, g is the gravitational acceleration, R is the radius of the cylinder cup, β_ω is the damping coefficient, and ω_0 is the natural frequency.

2.2. General Theory

Based on preceding assumptions, the Navier-Stokes equation can be simplified into:

$$\frac{\partial \mathbf{u}}{\partial t} + (\mathbf{u} \cdot \nabla)\mathbf{u} = -\nabla \frac{p}{\rho} + \mathbf{g} + \mathbf{a}, \qquad (1)$$

where $\mathbf{u}$ is the velocity field of the fluid, p is the pressure distribution, $\mathbf{g}$ is the gravitational acceleration, and $\mathbf{a}$ is an arbitrary acceleration depending on different situations. Introducing $\mathbf{u} = -\nabla\Phi$, where Φ is velocity potential, together with the continuity condition, the total governing equations for this problem are:

$$\begin{cases} \nabla^2\Phi = 0 & \text{Continuity condition} \\[2mm] \dfrac{\partial\Phi}{\partial n} = 0 & \text{Rigid wall condition} \\[2mm] \dfrac{\partial\Phi}{\partial t} = g\eta + \mathbf{r} \cdot \mathbf{a} & \text{Dynamic boundary condition} \\[2mm] \dfrac{\partial\Phi}{\partial z} = \dfrac{\partial\eta}{\partial t} & \text{Kinematic free-surface condition} \end{cases} \qquad (2)$$

where $\eta(x, y, t)$ is the surface equation, $\mathbf{r}$ is position vector.

It should be mentioned that dynamic boundary condition is derived from Eq. (1), under the assumption that the flow is irrotational, together with the use of $(\mathbf{u} \cdot \nabla)\mathbf{u} = \frac{1}{2}\nabla \mathbf{u}^2 - \mathbf{u} \times (\nabla \times \mathbf{u}) = \frac{1}{2}\nabla \mathbf{u}^2$.

Continuity condition illustrates that the water flow is continuum, in an everyday speech that the volume of the water body remains unchanged. The rigid wall condition tells that the normal velocity is zero at the cup wall. For different excitations, the dynamic boundary condition should be modified. In the following sections, there will be detailed discussions on specific acceleration $\mathbf{a}$.

In this work, a cylindrical cup is mainly used. From the first two equations in Eq. (2), the general solution in a cylindrical region can be read as:

$$\Phi(r,\theta,z,t) = \sum_{m=0}^{\infty} \sum_{n=1}^{\infty} [\alpha_{mn}(t)\cos m\theta + \beta_{mn}(t)\sin m\theta]$$
$$\times J_m(\lambda_{mn}r)\frac{\cosh[\lambda_{mn}(z+h)]}{\cosh\lambda_{mn}h}$$

(3)

Fig. 2.   Sketch of cylindrical cup.

where h and r are the height and radius of the water body (as shown in Fig. 2 respectively), J_m is the Bessel function of m order, $\lambda_{mn} = \zeta_{mn}/R$, and ζ_{mn} is the root of $\frac{\partial J_m(\lambda_{mn}r)}{\partial r}\big|_{r=R} = 0$. Some ζ_{mn} can be found in Table 1: For different specific motions, the general solution can be substituted

Table 1.   The values of ζ_{mn} with different m and n

	$n=1$	$n=2$	$n=3$	$n=4$	$n=5$
$m=0$	3.832	7.016	10.174	13.324	16.471
$m=1$	1.841	5.331	8.536	11.706	14.864
$m=2$	0	3.054	6.706	9.969	13.170
$m=3$	0	4.201	8.015	11.3459	14.586
$m=4$	0	5.318	9.282	12.681	15.964
$m=5$	0	6.416	10.520	13.987	17.313

in different dynamic boundary conditions together with kinematic free-surface boundary condition to find corresponding analytic solutions.

The indefinite amplitude function $\alpha(t)$ and $\beta(t)$ can be determined by the dynamic boundary condition for different shapes of the coffee cup.

Some comments should be conducted. In some trivial cases and typical vibrations, the symmetry can make the general solution readable. In this problem, as odd symmetry of horizontal vibration leads one of α and β to be zero, or, as the even symmetry of the vertical vibration causes $\alpha = \beta$, the amplitude function α or β only controls the absolute magnitude of surface height derivation from the equilibrium position depending on the time instead of the relative magnitude. In such symmetries, the vibration patterns remain unchanged at specific frequency ω.

The derived quantities of velocity **u**, pressure p and surface equation η, are listed as follow:

$$\mathbf{u} = -\nabla\Phi$$
$$\eta = -\frac{\partial\Phi}{\partial t} - \mathbf{r}\cdot\mathbf{a} \qquad (4)$$
$$p = \rho\frac{\partial\Phi}{\partial t}$$

These quantities will be used to analyse water motion in each stage.

3. Experimental Process and Data Analysis

A cup with different volumes of water is settled on a vibration platform. Two drops of ink are added to the water to facilitate data extraction. The signal generator() is used to drive the stage. The experiment process is recorded by the camera() and analysed by using the software "Tracker" to get the velocity and acceleration of characteristic points. To analyse the frequency, we conduct the discrete Fourier transformation on the data in the time domain (Fig. 3).

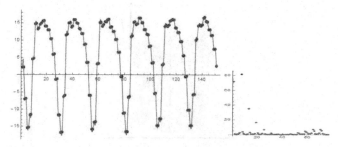

(a) Data points of displacement for a vibration experiment.

(b) After discrete fourier transformation.

Fig. 3.   Frequency analysis.

4. Detailed Analysis

In this section, we will analyze different stages (Fig. 1) in a detailed way theoretically and experimentally.

4.1. *First Stage: Uniform Acceleration*

In the first stage, the acceleration effect dominates. We will investigate the trivial situation, uniform acceleration, setting the constant acceleration along x-axis, thus,

$$-\frac{\partial \Phi}{\partial t} = g\eta + r\cos\theta \cdot a \tag{5}$$

The surface equation can be derived directly from Eq. (5):

$$\eta = -\frac{1}{g}\frac{\partial \Phi_0}{\partial t} - \frac{a}{g}r\cos\theta \tag{6}$$

The term of Φ_0 depends on the preceding proceed. Intuitively, Φ_0 is the addition of a slope surface and the initial surface starting from the end of the preceding motion. Fig. 4 is the illustration of water surface in the first stage. Obviously, if the coffee is too full, or the acceleration is too large, the coffee will be more likely to spill out of the cup.

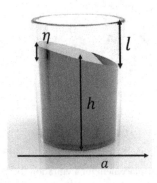

Fig. 4.   Uniform acceleration.

4.2. Second Stage: Steady Motion

While in the steady motion, as the acceleration is a small quantity, we assume that the dominant factor is the periodical perturbation which can be decomposed into 3 types of the vibration, namely, horizontal vibration, vertical vibration and the pitching vibration. Next, we will investigate these types of vibration separately.

4.2.1. Vertical Vibration

In this situation, the cup undergoes a vertical periodical acceleration. Hence, the dynamic boundary condition is

$$-\frac{\partial \Phi}{\partial t} = (g + \epsilon \cos (\omega t)) \, \eta, \tag{7}$$

where ϵ is the amplitude of the vertical excitation and ω is the circular frequency of the excitation.

Combining Eq. (7) with kinematic free-surface condition, and substituting the general solution into it, we have the amplitude equation for $\alpha(t), \beta(t)$, where $\alpha = \beta$:

$$\ddot{\alpha}_{mn}(\tau) + (\delta + \varepsilon \cos (2\tau)) \, \alpha_{mn}(\tau) = 0,$$
$$\tau = \frac{\omega}{2} t, \delta = \frac{\xi_{mn} \tanh\left(\frac{\xi_{mn} h}{R}\right)}{R} g, \varepsilon = \frac{\xi_{mn} \tanh\left(\frac{\xi_{mn} h}{R}\right)}{R} \epsilon \tag{8}$$

This is a typical Mathieu's equation, demonstrating that the natural frequency for this system is $\omega_{mn}^2 = \frac{g \xi_{mn}}{R} \tanh (\xi_{mn} h / R)$. As said in the introduction section, the frequency of human excitation ranges from 1hz to 3.5hz. Thus, Fig. 5 illustrates that $m = 1$ mode is more likely to happen.

Mathieu equation predicts the existence of the subharmonic resonance which the frequency of excitation is double of the natural frequency. Fig. 6 is a phase diagram containing harmonic and subharmonic region.

We conduct experiments to analyze the vertical vibration (Fig. 7(a)). Changing the height of water body and recording the resonance frequency, the relationship between resonance frequency and the geometry parameters have been found (Fig. 7(b)).

In Fig. 7(b), the blue lines are the theoretical curves, and the red points are experimental data. It is clear that there exist the harmonic and subharmonic phenomena ($\omega = 2\omega_0$). The theoretical curves have a good agreement with the data.

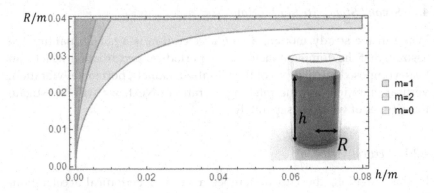

Fig. 5.   First 3 orders of resonance patterns.

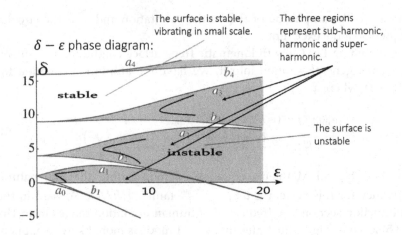

Fig. 6.   Phase diagram of Mathieu equation.

As the frequency of human motion is about 2Hz, which is much less than the frequency of the subharmonic resonance, the vertical vibration is hard to appear.

4.2.2. *Horizontal Vibration*

The dynamic boundary condition for the horizontal vibration is similar to the uniform acceleration, while the acceleration is not a constant but a periodical function, namely, $\ddot{X} = X_0\Omega^2 \cos(\Omega t)$. And the boundary

 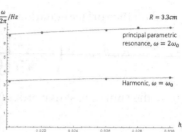

(a) Apparatus for vertical vibration.   (b) Data of vertical vibration, subharmonic can be found.

Fig. 7. Experiment verification of vertical vibration.

condition is

$$g\eta - \frac{\partial \Phi}{\partial t} + \ddot{X}r\cos(\theta) = 0, z = \eta(r,\theta,t), \tag{9}$$

Combining Eq. (9) with the kinematic free-surface condition, we have

$$g\frac{\partial \Phi}{\partial z} + \frac{\partial^2 \Phi}{\partial t^2} = \frac{d^3 X}{dt^3}r\cos\theta \tag{10}$$

As the general solution is in the form of infinite Bessel series, r in the RHS of Eq. (10) should be changed into the form of Fourier-Bessel series:

$$r = \sum_{n=1}^{\infty} \frac{2R}{(\lambda_{1n}^2 R^2 - 1) J_1(\lambda_{1n}R)} J_1(\lambda_{1n}r) \tag{11}$$

Substituting Eq. (11) and general solution Eq. (3) into Eq. (10), we have two equations for the amplitude functions:

$$\ddot{\alpha}_{1n}(t) + \omega_{1n}^2 \alpha_{1n}(t) = \frac{d^3 X}{dt^3} \frac{F_n}{\cosh k_{1n}h} \tag{12}$$
$$\ddot{\beta}_{1n}(t) + \omega_{1n}^2 \beta_{1n}(t) = 0$$

There only exist Bessel functions of order $m = 1$, indicating that the amplitude functions of order $m \neq 1$ vanish. The total velocity potential is:

$$\Phi = -X_0 \Omega \cos\theta \cos\Omega t \cdot$$
$$\left(r + \sum_{n=1}^{\infty} \left[\frac{2R}{(\xi_{1n}^2 - 1)} \frac{\Omega^2}{\omega_{1n}^2 - \Omega^2} \frac{J_1(\xi_{1n}r/R)}{J_1(\xi_{1n})} \frac{\cosh[\xi_{1n}(z+h)/R]}{\cosh(\xi_{1n}h/R)} \right] \right) \tag{13}$$

And the natural frequency for horizontal vibration only has one order, namely, $\omega_{1n}^2 = \frac{g\xi_{1n}}{R} \tanh(\xi_{1n}h/R)$.

Based on the surface equation which is

$$\eta = \frac{X_0\Omega^2}{g}\cos\theta\sin\Omega t\left(r + \sum_{n=1}^{\infty}\left[\frac{2R}{(\xi_{1n}^2-1)}\frac{\Omega^3}{\omega_{1n}^2-\Omega^2}\frac{J_1(\xi_{1n}r/R)}{J_1(\xi_{1n})}\right]\right),$$
(14)

we draw the sum of first five order in Fig. 8.

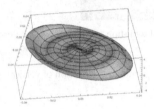

Fig. 8.   Surface of horizontal vibration.

By using Eq. (4), we can find the pressure distribution (Fig. 9) for the horizontal vibration,

$$p = \rho\frac{\partial\Phi}{\partial t}$$
$$= -\rho X_0\Omega^2\cos\theta\sin\Omega t$$
$$\left(r + \sum_{n=1}^{\infty}\left[\frac{2R}{(\xi_{1n}^2-1)}\frac{\Omega^2}{\omega_{1n}^2-\Omega^2}\frac{J_1(\xi_{1n}r/R)}{J_1(\xi_{1n})}\frac{\cosh[\xi_{1n}(z+h)/R]}{\cosh(\xi_{1n}h/R)}\right]\right)$$

The velocity field is $\mathbf{u} = -\nabla\Phi$. The velocity field diagram (Fig. 10) shows

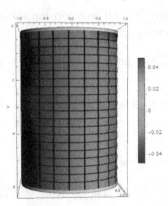

Fig. 9.   Pressure distribution on the cup walls.

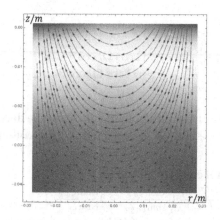

Fig. 10.   Velocity field in the middle section.

that the velocity is larger at the top, and the open streamline indicates the water will be easier to get out.

Fig. 11(a) is the experimental set-up for the horizontal vibration. And the theoretical curve has a good agreement with the data (Fig. 11(b)).

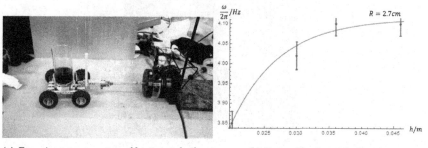

(a) Experiment apparatus of horizontal vibra-   (b) Data of horizontal vibration
tion.

Fig. 11.   Experiment verification of horizontal vibration.

4.2.3. Pitching Vibration

Pitching vibration means that the cup sways along a horizontal axis. The excitation can be described by the angle Ωt between the cup and the vertical direction, $\psi(t) = \psi_0 sin(\Omega t)$. The boundary conditions for pitching vibration are:

$$\nabla^2 \Phi = 0$$
$$\frac{\partial \Phi}{\partial r} = -\psi_0 z \Omega \cos \Omega t \cos \theta \,|_{r=R}$$
$$-\frac{\partial \Phi}{\partial z} = -\psi_0 r \Omega \cos \Omega t \cos \theta \,|_{z=-h/2} \tag{15}$$
$$\frac{\partial^2 \Phi}{\partial t^2} + g\frac{\partial \Phi}{\partial z} = 0 \,|_{z=h/2}$$

Due to the linearity of velocity potential Φ, Φ can be decomposed to fit the boundary conditions, and we have a lengthy total potential:[4]

$$\Phi = -\Omega\psi_0 \cos \Omega t \cos \theta \, (rz + A + B)$$
$$A = \sum_{n=1}^{\infty} \frac{R^2}{(\xi_n^2-1)} \frac{J_1\left(\xi_n \frac{r}{R}\right)}{J_1(\xi_n)} \left(\frac{\cosh \xi_n \frac{z}{R}}{\sinh \xi_n \frac{h}{2R}} - 3\frac{\sinh \xi_n \frac{z}{R}}{\cosh \xi_n \frac{h}{2R}} \right)$$
$$B = \sum_{n=1}^{\infty} \frac{\Omega^2}{\omega_n^2 - \Omega^2} \frac{R^2 \cosh \frac{\xi_n}{R}\left(z+\frac{h}{2}\right)}{(\xi_n^2-1)} \left(\frac{\left(\xi_n\frac{h}{R}\right)+\coth\left(\xi_n\frac{h}{2R}\right)-3\tanh\left(\xi_n\frac{h}{2R}\right)}{\xi_n\cosh\left(\xi_n\frac{h}{2R}\right)} \right) \frac{J_1\left(\xi_n\frac{r}{R}\right)}{J_1(\xi_n)}$$

$$\tag{16}$$

Although it is too lengthy to find the feature, we can draw the velocity field in Fig. 12. The diagram demonstrates that the streamlines are extremely

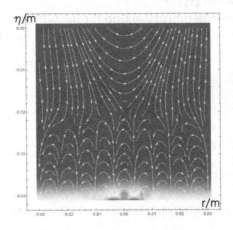

Fig. 12.   Velocity field for pitching vibration in the middle section.

sloped at the surface of the water body, hinting that the surface water tends to get out, while the water at the bottom tends to convect.

We conduct the experiment by the apparatus shown in Fig.13. It appears that when the frequency of excitation Ω is close to the natural frequency, the water will resonate and get out, which is consistent with the solution.

Fig. 13.   Experiment apparatus of the pitching vibration.

Actually, in the 2nd stage, the motion of the water in the cup is composed of the three main modes, namely, horizontal vibration, vertical vibration, pitching vibration.

4.3. Third Stage: Uniform Deceleration

In the 3rd stage, where the deceleration effect dominates, the dynamic boundary condition can be written as

$$-\frac{\partial \Phi}{\partial t} = g\eta + r\cos\theta \cdot a \tag{17}$$

The surface equation can be deduced according to the dynamic boundary condition, which is,

$$\eta = -\frac{1}{g}\frac{\partial \Phi_0}{\partial t} - \frac{a}{g}r\cos\theta \tag{18}$$

where the Φ_0 depends on the final state of the 2nd stage. In this stage, the water experiences a combination of slope surface and damping vibration. If one walks faster in the second stage and stops suddenly, the water is likelier to spill out in the stage.

4.4. Fourth Stage: Free Vibration

Free vibration is the most trivial but significant situation, it can be regarded as the transition of discontinuous process. However, in our analysis, the free vibration dominates when the human motion ceased, which is the 4th stage.

The dynamic boundary condition is simple, $-\frac{\partial \Phi}{\partial t} = g\eta, z = \eta\left(x^i, t\right)$. Combining the dynamic boundary condition with kinematic free-surface condition, we have

$$\frac{\partial^2 \Phi}{\partial t^2} + g\frac{\partial \Phi}{\partial z} = 0 \tag{19}$$

Substituting the general solution into it, we have the equations for the amplitude functions:

$$\begin{aligned}\ddot{\alpha}_{mn}(t) + g\lambda_{mn}\tanh\left(\lambda_{mn}h\right)\alpha_{mn}(t) = 0 \\ \ddot{\beta}_{mn}(t) + g\lambda_{mn}\tanh\left(\lambda_{mn}h\right)\beta_{mn}(t) = 0\end{aligned} \tag{20}$$

From Eq. (20), we find the natural frequency is $\omega_{mn} = \sqrt{\frac{g\zeta_{mn}}{R}}\tanh\left(\zeta_{mn}h/R\right)$, here m can take any integer larger than 0, n can take any integer larger than 1.

5. Conclusion

In the conclusion, we will return to the 3 stages in the introduction section and Fig. 1. As the preceding analysis suggests, in the first stage, which is assumed to be a uniform acceleration process, the water surface is slope, and the larger acceleration, the higher water surface. At the point it comes to the second stage, one must be cautious, since in this discontinuous point here, and the water is more likely to spill out. In the steady motion stage, horizontal vibration appears most likely, and vertical vibration appears less likely for the subharmonic phenomenon is hard to reach. The pitching vibration appears the least likely, unless you hold a mug, means, hold the handle of the cup. When it comes to the third stage, you also need to be cautious, for the addition of the slope surface and preceding vibration surface is probably higher than the edge of the cup.

6. Discussion

6.1. *The Shapes of the Cup*

To find the different effects between the cylindrical region and the rectangle region, we have to solve the Navier-Stokes equation in the rectangle

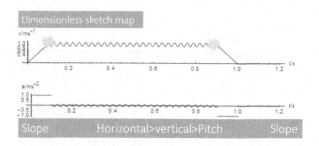

Fig. 14.   Different stages in human motion. The yellow stars mark the point where the water is most likely to spill out. In the steady motion stage, the likelihood of water being spilt out due to the effect of the horizontal vibration is larger than that of the vertical and the pitching one.

region, the governing equations which are similar to Eq. (2) are:

$$\nabla^2 \Phi = 0$$
$$\partial_x \Phi = 0, x = \pm \tfrac{L}{2} \partial_z \Phi = 0, z = -h$$
$$g\eta - \partial_t \Phi + a_x x = 0, z = \eta$$
$$-\partial_z \Phi = \partial_t \eta, z = \eta$$

(21)

Combining the first equation of Eq. (21) with the boundary condition, we can find the general solution:

$$\Phi = \sum_{m=1}^{\infty} \left(\alpha_m(t) \cos(k_m x) + \beta_m(t) \sin(k_m x) \right) \cosh(k_m(z+h))$$

where $k_m = \frac{2(m-1)}{L}\pi$ or $k_m = \frac{2m}{L}\pi$. Combining the dynamic boundary condition and kinematic free-surface condition, we get:

$$\dot{a}_x x = \frac{\partial^2 \Phi}{\partial t^2} + g \frac{\partial \Phi}{\partial z}$$

(22)

Rewriting x as Fourier series, and substituting the general solution into the Eq. (22), we have

$$\cosh\left(\frac{(2n+1)\,\pi h}{L} \right) \frac{4L}{\pi^2} \frac{(-1)^n}{(2n+1)^2} \dot{a}_x$$
$$= g\frac{(2n+1)}{L}\pi \left(\frac{(2n+1)\,\pi h}{L} \right) \alpha(t) + \ddot{\alpha}(t)$$

Thus the natural frequency is $\omega^2 = g\frac{(2n+1)\pi}{L} \tanh\left(\frac{(2n+1)\pi}{L} h \right)$. If we put the same volume of water in the cylindrical and rectangle cup respectively, the natural frequency of rectangle cup is closer to the frequency of human motion, as illustrated in Fig. 15, thus, the water in the rectangle cup is more likely to spill out.

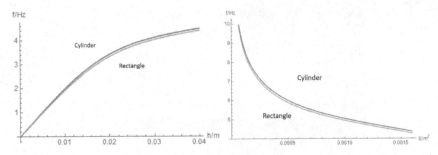

(a) Natural frequency vs. the height of water (b) Natural frequency vs. the section area of body.   water body.

Fig. 15.   Comparison between the effects of the cylindrical cup and the rectangle cup.

6.2. *High Frequency Noise*

When we decompose the actual motion into Fourier series, there will have high frequency components. These components contribute to the asymmetry of the pattern. Eventually, the motion evolves to rotating or splashing.

References

1. E. Widmayer Jr and J. R. Reese, Moment of inertia and damping of fluid in tanks undergoing pitching oscillations (1953).
2. H. Takahara, K. Kimura, and M. Sakata, Frequency response of sloshing in a circular cylindrical tank subjected to pitching excitation, *Trans JSME C.* **60**(572), 1210–1216 (1994).
3. K. Brown, Laboratory test of fuel sloshing, *Douglas Aircraft, CA, Rep Dev.* **782**, 18 (1954).
4. R. A. Ibrahim, *Liquid sloshing dynamics: theory and applications.* Cambridge University Press (2005).

Author Index

Subject Index

velocity potential, 151
vertical vibration, 153
vibration of the plate, 80

Printed in the United States
By Bookmasters